Guide for the

Local Calibration of the Mechanistic-Empirical Pavement Design Guide

November 2010

AMERICAN ASSOCIATION OF STATE HIGHWAY AND TRANSPORTATION OFFICIALS

AASHTO

1914

Publication Code: LCG-1
ISBN: 978-1-56051-449-7

American Association of State Highway and Transportation Officials

Executive Committee 2009/2010

President: Larry L. "Butch" Brown, Sr., Mississippi
Vice President: Susan Martinovich, Nevada
Secretary-Treasurer: Carlos Braceras, Utah

REGIONAL REPRESENTATIVES

REGION I

Joseph Marie, Connecticut, One-Year Term
Gabe Klein, District of Columbia, Two-Year Term

REGION II

Dan Flowers, Arkansas, One-Year Term
Mike Hancock, Kentucky, Two-Year Term

REGION III

Nancy J. Richardson, Iowa, One-Year Term
Thomas K. Sorel, Minnesota, Two-Year Term

REGION IV

Paula Hammond, Washington, One-Year Term
Amadeo Saenz, Jr., Texas, Two-Year Term

NON-VOTING MEMBERS

Immediate Past President: Allen Biehler, Pennsylvania
AASHTO Executive Director: John Horsley, Washington, DC

Joint Technical Committee on Pavements 2008/2009

Chair: David Nichols, Missouri
Vice Chair: Judith Corley-Lay, North Carolina
Secretary: Pete Stephanos, FHWA

REGION I
Delaware
Robin Davis (Design)

Maryland
Tim Smith (Materials)

New York
Wes Yang (Design)
Vacant (Design)
Vacant Design)

REGION II
Alabama
Larry Lockett (Materials)

Arkansas
Phillip McConnell (Design)

Louisiana
Jeff Lambert (Design)

North Carolina
Judith Corley-Lay (Vice Chair)

South Carolina
Andy Johnson (Design)
Kentucky
Paul Looney (Design)

REGION III
Kansas
Andy Gisi (Design)

Minnesota
Curt Turgeon (Materials)

Missouri
Jay F. Bledsoe (Design)

Missouri
David Nichols (Chair)

Ohio
Aric Morse (Design)

Iowa
Chris Brakke (Design)

REGION IV
California
Bill Farnbach (Design)

Colorado
Richard Zamora (Design)

Oklahoma
Jeff Dean (Design)

Wyoming
Rick Harvey (Materials)
Vacant (Design)

Preface

This guide is to provide guidance to calibrate the *Mechanistic-Empirical Pavement Design Guide* (MEPDG) software to local conditions, policies, and materials and to conduct the local calibration process. The guide does not provide guidance for determining the inputs and running the MEPDG software. A separate document, the *Mechanistic-Empirical Pavement Design Guide—A Manual of Practice*, provides guidance for using the MEPDG software to analyze and design new pavements and rehabilitation strategies. The *Manual of Practice* is referenced throughout this guide.

Version 1.0 of the MEPDG software is currently available. It should be noted that version 2.0 of the MEPDG software is in the process of being developed. Version 2.0 may include different transfer functions for selected distresses based on the results and recommendations from other on-going NCHRP projects. If any of the transfer functions are revised, the *Guide for Local Calibration* and the *Mechanistic-Empirical Pavement Design Guide—A Manual of Practice* for the MEPDG software may need to be revised accordingly.

Table of Contents

List of Figures

Attachments

Attachments

List of Tables

1.0 Introduction

The overall objective of the *Mechanistic-Empirical Pavement Design Guide* (MEPDG) is to provide the highway community with a state-of-the-practice tool for the design of new and rehabilitated pavement structures, based on mechanistic-empirical (M-E) principles. This means that the design procedure calculates pavement responses (stresses, strains, and deflections) and uses those responses to compute incremental damage over time. The procedure empirically relates the cumulative damage to observed pavement distresses. This M-E based procedure is shown in flowchart form in Figure 1-1. "MEPDG," as used in this guide, refers to the documentation and software package (NCHRP 2007).

Pavement distress prediction models, or transfer functions, are the key components of any M-E design and analysis procedure. The accuracy of performance prediction models depends on an effective process of calibration and subsequent validation with independent data sets. Pavement engineers gain confidence in the procedure by seeing an acceptable correlation between observed levels of distress in the field and those levels predicted with the performance model or transfer function. The validation of the performance prediction model is a mandatory step in their development to establish confidence in the design and analysis procedure and facilitate its acceptance and use. It is also necessary to establish the design reliability procedure. It is essential that distress prediction models be properly calibrated prior to adopting and using them for design purposes.

The term calibration refers to the mathematical process through which the total error (often termed residual) or difference between observed and predicted values of distress is minimized. The term validation refers to the process to confirm that the calibrated model can produce robust and accurate predictions for cases other than those used for model calibration. A successful validation process requires that the bias and precision statistics of the model for the validation data set be similar to those obtained during calibration. This calibration-validation process is critical for potential users to have confidence in the design procedure.

All performance models in the MEPDG were calibrated on a global level to observed field performance over a representative sample of pavement test sites throughout North America. The Long Term Pavement Performance (LTPP) test sections were used extensively in the calibration process, because of the consistency in the monitored data over time and the diversity of test sections spread throughout North America. Other experimental test sections were also included such as

MnRoad and Vandalia. However, policies on pavement preservation and maintenance, construction and material specifications, and materials vary across the United States and are not considered directly in the MEDPG. These factors can be considered indirectly through the local calibration parameters included in the MEPDG. The purpose of this guide is to provide guidance in calibrating the MEPDG to local conditions and materials that may not have been included in the global calibration process.

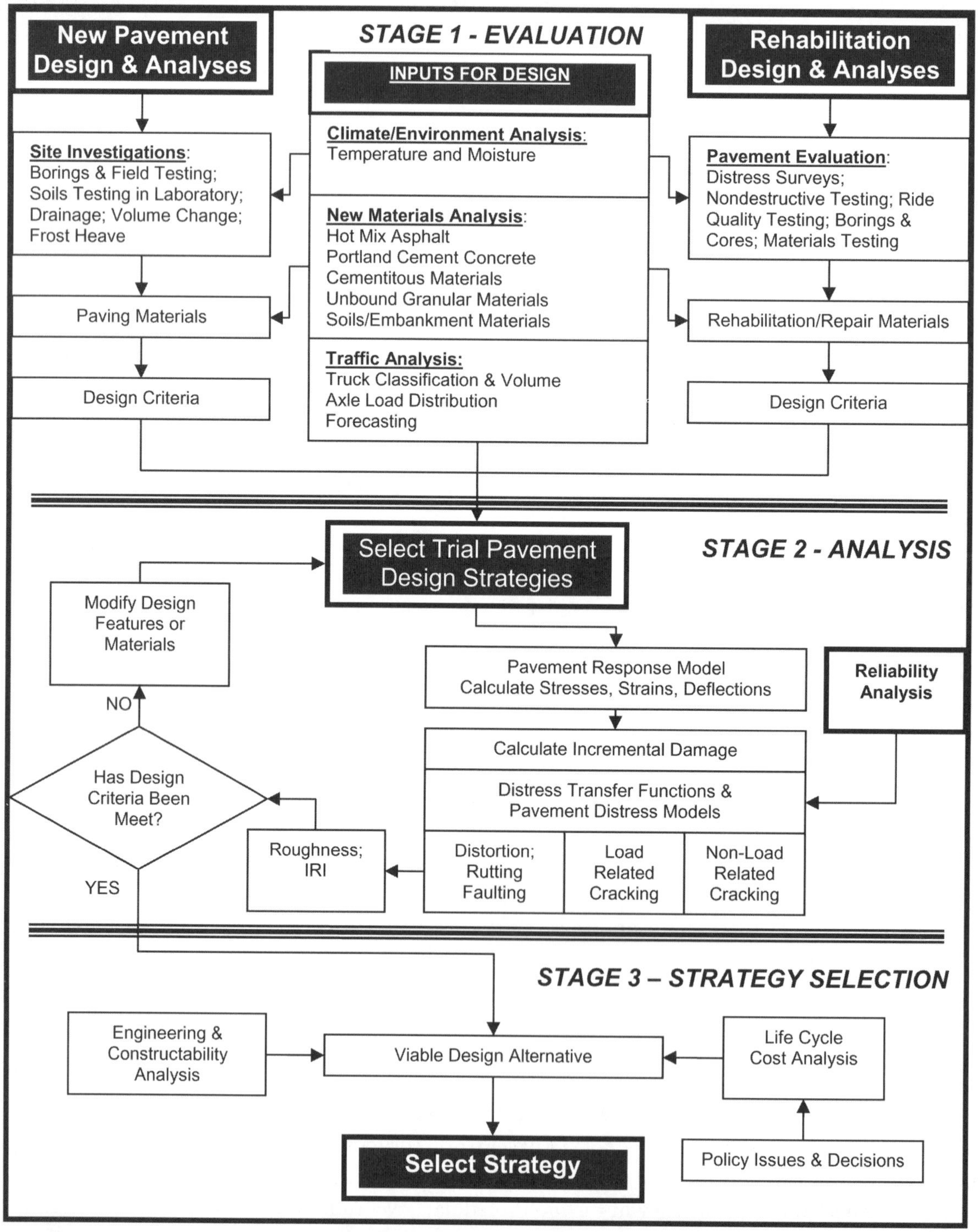

Figure 1-1. Conceptual Flow Chart of the Three-Stage Design/Analysis Process for the MEPDG

2.0 Terminology and Definition of Terms

This section provides the definition of selected terms that are used in the MEPDG local calibration process.

2.1 Statistical Terms

Accuracy—The exactness of a prediction to the observed or "actual" value. The concept of accuracy encompasses both precision and bias.

Bias—An effect that prevents predictions from simulating "real world" observations by systematically distorting it, as distinct from a random error that may distort on any one occasion but balances out on the average. A prediction model that is "biased" is significantly over- or under-predicting observed distress or roughness (as measured by the International Roughness Index [IRI]).

Calibration—A systematic process to eliminate any bias and minimize the residual errors between observed or measured results from the real world (e.g., the measured mean rut depth in a pavement section) and predicted results from the model (e.g., predicted mean rut depth from a permanent deformation model). This is accomplished by modifying empirical calibration parameters or transfer functions in the model to minimize the differences between the predicted and observed results. These calibration parameters are necessary to compensate for model simplification and limitations in simulating actual pavement and material behavior.

Model, Mathematical—A model that is derived from fundamental engineering principles that represent exact, error-free assumed relationships among the variables. The JULEA program is the mathematical or structural response model used for flexible pavements to calculate pavement responses (deflections, stresses, and strains), while the ISLAB2000 program is used for rigid pavements. A stress dependent finite element program is also available for flexible pavement analyses using Input Level 1 for unbound materials, but that model is intended for research purposes only. The Integrated Climatic Model (ICM) is also considered a mathematical model within the MEPDG.

Model, Statistical—A model that is derived from data that are subject to various types of observational, experimental, and measurement errors. The statistical models in the MEPDG include the distress transfer functions and the IRI regression equations. The time-dependent material property models for hot mix asphalt (HMA) and Portland cement concrete (PCC) are also regression or statistical relationships. These models, however, are assumed to be correct in the MEPDG model formulation or computational methodology. Adjustments to the coefficients of these relationships are not permitted within Version 1.0 of the MEPDG.

Model, Simulation or Prediction—Prediction models take two related forms. First the real-world system under investigation is approximated by a conceptual model. A conceptual model is a series of mathematical and logical relationships concerning the components and the structure of the system. Then the conceptual model is coded into a computer-recognizable form, the operation model, which is an approximate representation of the real-world system. The MEPDG operation models combine mathematical and statistical models.

Precision—The ability of a model to give repeated estimates that correlate strongly with the observed values. They may be consistently higher or lower but they correlate strongly with observed values.

Residual Error—The difference between the observed or measured and predicted distress and IRI values (e.g., measured minus predicted values). The residuals contain the available information about how well the model predicts the observed distress and IRI.

Standard Error of the Estimate (s_e)—The standard deviation of the residual errors for the pavement sections included in the calibration data set for each prediction model. The standard error is usually obtained by taking the positive square root of the variance of the statistic.

Validation—A systematic process that re-examines the recalibrated model to determine if the desired accuracy exists between the calibrated model and an independent set of observed data. The calibrated model required inputs such as the pavement structure, traffic loading, and environmental data. The simulation model must predict results (e.g., rutting, fatigue cracking) that are reasonably close to those observed in the field. Separate and independent data sets should be used for calibration and validation. Assuming that the calibrated models are successfully validated, the models can then be recalibrated using the two combined data sets without the need for additional validation to provide a better estimate of the residual error.

Verification—Verification of a model examines whether the operational model correctly represents the conceptual model that has been formulated. Verification can be achieved for simple models by comparing the model predictions (e.g., stress) against other analytical solutions for specific cases. Verification can also be accomplished by entering typical materials, structural, environmental, and traffic data into the distress and performance models, and then determining through parameter studies whether the program operates rationally and provides outputs that meet the criterion of engineering reasonableness. If this criterion is not met, the computer code maybe erroneous or the conceptual model may be unsatisfactory. In either case, these problems must be remedied before the model enhancement process continues. No field data are needed in either of the verification approaches described. Verification is primarily intended to confirm the internal consistency or reasonableness of the model. The issue of how well the model predicts reality is addressed during calibration and validation. As such, verification of the MEPDG prediction models is not included within this *Local Calibration Guide*.

2.2 MEPDG Calibration Terms

Calibration Factors—Two calibration factors are used in the MEPDG: global and local. These calibration factors are adjustments applied to the coefficients and/or exponents of the transfer function to eliminate bias between the predicted and measured pavement distresses and IRI. The combination of calibration factors (coefficients and exponents for the different distress prediction equations) can also be used to minimize the standard error of the prediction equation. The standard error of the estimate (s_e) measures the amount of dispersion of the data points around the line of equality between the observed and predicted values. The *MEPDG Manual of Practice* presents the calibration parameters for each distress prediction model included in the MEPDG (AASHTO, 2008).

Damage, Incremental—Incremental damage (ΔDI) is a ratio defined by the actual number of axle load applications (n) for a specified axle load and type within an interval of time divided by the allowable number of axle load applications (N) to some design criteria defined for the same axle load and type for the conditions that exist within the same specific period of time. The incremental damage indices are summed to determine the cumulative damage index over time.

Reliability—The probability that the predicted performance indicator of the trial design will not exceed the design criteria within the design-analysis period. The design reliability (R) is similar, in concept, to that in the current AASHTO *Pavement Design Guide*—the probability that the pavement will not exceed specific failure criteria over the design traffic. For example, a design reliability of 90 percent represents the probability (9 out of 10 projects) that the mean faulting for the project will not exceed the faulting criteria over the analysis period. The reliability of a particular design analyzed by the MEPDG is dependent on the standard errors of the transfer functions. The *User Manual* provides a more complete discussion on reliability and its use in the MEPDG for analyzing a trial design or pavement structure (AASHTO, 2008). However, a reliability level of 50 percent should always be used for predicting distresses to confirm or adjust the local calibration coefficients in accordance with this manual. This is further explained in Step 7 of Section 6 (Step-by-Step Procedure for Local Calibration).

Structural Response Model—*See model, mathmatical.*

Transfer Function—*See model, statistical.*

2.3 Hierarchical Input Level Terms

The hierarchical input level included in the MEPDG is an input scheme that is used to categorize the designer's knowledge of the input parameter. Three levels are available for determining the input values for most of the material and traffic parameters. The *MEPDG Manual of Practice* provides more detailed discussion on the purpose, use, and selection of the hierarchical input level for pavement design (AASHTO, 2008). The following defines each hierarchical input level that can be used by the designer:

Input Level 1—Input parameter is measured directly; it is site- or project-specific. This level represents the greatest knowledge about the input parameter for a specific project but has the highest testing and data collection costs to determine the input value. Level 1 should be used for pavement designs having unusual site features, materials, or traffic conditions that are outside the inference-space used to develop the correlations and defaults included for Input Levels 2 and 3.

Input Level 2—Input parameter is estimated from correlations or regression equations. The input value is calculated from other site specific data or parameters that are less costly and/or easier to measure. Input Level 2 can also represent measured regional values that are not project-specific.

Input Level 3—Input parameter is based on "best-estimated" or default values. Level 3 inputs are based on global or regional default values—the median value from a group of data with similar characteristics. This input level has the least knowledge about the input parameter for the specific project but has the lowest testing and data collection costs.

2.4 Distress or Performance Indicator Terms

This subsection provides a definition of each distress and performance indicator predicted by the MEPDG. It also provides the standard errors of the estimate for each transfer function that are considered reasonable, and are similar to the values included in the *MEPDG Manual of Practice* (AASHTO, 2008). A reasonable standard error of the estimate, however, will be dependent on the design or threshold value used by the agency in their day-to-day management practices.

Hot Mix Asphalt (HMA)-Surfaced Pavements

Alligator Cracking—A form of fatigue or load-related cracking defined as a series of interconnected cracks (characteristically with a "chicken wire/alligator" pattern) that initiate at the bottom of the HMA layers. Alligator cracks initially show up as multiple short, longitudinal, or transverse cracks in the wheel path that become interconnected laterally with continued truck loadings. Alligator cracking is calculated as a percent of total lane area in the MEPDG. The MEPDG does not predict the severity of alligator cracking, but includes low, medium, and high in the definition. A reasonable standard error of the estimate for alligator or bottom-up cracking is seven percent.

Longitudinal Cracking—A form of fatigue or load-related cracking that occurs within the wheel path, defined as cracks parallel to the pavement centerline. Longitudinal cracks initiate at the surface of the HMA pavement and initially show up as short longitudinal cracks that become connected longitudinally with continued truck loadings. Raveling or crack deterioration can occur along the edges of these cracks but they do not form an alligator cracking pattern defined above. The unit of longitudinal cracking calculated by the MEPDG is feet per mile (meters per kilometer). The MEPDG does not predict severity of the longitudinal cracks, but includes low, medium, and high in the definition. A reasonable standard error of the estimate for longitudinal or top-down cracking is 600 ft/mi.

Unless an agency cuts cores or trenches through the HMA surface to confirm where the cracks initiated, it is recommended that the local calibration refinement be confined to total cracking that combines alligator and longitudinal cracks. To combine percent total lane area fatigue cracks with linear or longitudinal fatigue cracks, the total length of longitudinal cracks should be multiplied by 1 ft and that area divided by the total lane area. When an agency decides to combine alligator and longitudinal cracks, the alligator transfer function should be the one used in the local calibration process for determining the local calibration values. If an agency recovers cores or cuts trenches, but cannot determine where the cracks initiated, it is recommended that the agency assume all cracks initiated at the bottom of the HMA layer.

Reflective Cracking—Fatigue cracks in HMA overlays of flexible pavements and of semi-rigid and composite pavements, plus transverse cracks that occur over transverse cracks and joints and cracks in jointed PCC pavements. The unit of reflective cracking calculated by the MEPDG is feet per mi (meters per kilometer). The MEPDG does not predict the severity of reflective cracks but includes low, medium, and high in the definition. Unless an agency cuts cores or trenches through the HMA overlay of flexible pavements to confirm reflective cracks, it is recommended that the local calibration refinement be confined to total cracking of HMA overlays. In this case, all surface cracks in the wheel path (reflective, alligator, and longitudinal cracks) should be combined, using the recommendation for longitudinal cracking listed above. If all cracks are combined, the alligator and reflection cracking transfer functions can be used in the local calibration process.

Rutting or Rut Depth—A longitudinal surface depression in the wheel path resulting from plastic or permanent deformation in each pavement layer. The rut depth is representative of the maximum vertical difference in elevation between the transverse profile of the HMA surface and a wire-line across the lane width. The unit of rutting calculated by the MEPDG is inches (millimeters). A reasonable standard error of the estimate for total rutting is 0.10 in. The MEPDG also computes the rut depths within the HMA, unbound aggregate layers, and foundation. Unless an agency cuts trenches through pavement sections, however, it is recommended that the calibration refinement be confined to the total rut depth predicted with the MEPDG.

Transverse Cracking—Non-wheel load-related cracking that is predominately perpendicular to the pavement centerline and caused by low temperatures or thermal cycling. The unit of transverse cracking calculated by the MEPDG is feet per mile (meters per kilometer) or spacing of transverse cracks in feet. The MEPDG does not predict the severity of transverse cracks but includes low, medium, and high in the definition. A reasonable standard error of the estimate for transverse cracking is 250 ft/mi.

Portland Cement Concrete (PCC)-Surfaced Pavements

Faulting, Mean Transverse Joint (Jointed Plain Concrete Pavement [JPCP])—Transverse joint faulting is the differential elevation across the joint measured approximately 1 ft from the slab edge (longitudinal joint for a conventional lane width), or from the rightmost lane paint stripe for a widened slab. Since joint faulting varies significantly from joint to joint, the mean faulting of all transverse joints in a pavement section is the parameter predicted by the MEPDG. A reasonable standard error of the estimate for faulting is 0.05 in.

Faulting is an important deterioration mechanism of JPCP because of its impact on ride quality. Transverse joint faulting is the result of a combination of repeated applications of moving heavy axle loads, poor load transfer across the joint, free moisture beneath the PCC slab, erosion of the supporting base/subbase, subgrade, or shoulder base material, and upward curling of the slab.

Punchouts, Continuously Reinforced Concrete Pavement (CRCP)—When truck axles pass along near the longitudinal edge of the slab between two closely spaced transverse cracks, a high-tensile stress occurs at the top of the slab, some distance from the edge (typically 48 in. from the edge), transversely across the pavement. This stress increases greatly when there is loss of load transfer across the transverse cracks or loss of support along the edge of the slab. Repeated loading of heavy axles results in fatigue damage at the top of the slab, which results first in micro-cracks that initiate at the transverse crack and propagate longitudinally across the slab to the other transverse crack resulting in a punchout. The punchouts in CRCP are predicted considering the loss of crack load transfer efficiency (LTE) and erosion along the edge of the slab over the design life, and the effects of permanent and transitory moisture and temperature gradients. The transverse crack width is the most critical factor affecting LTE and, therefore, punchout development. Only medium- and high-severity punchouts, as defined by LTPP (FHWA, 2003), are included in the MEPDG model global calibration. A reasonable standard error of the estimate for the number of punchouts is four per mile.

Transverse Cracking, Bottom-Up (JPCP)—When the truck axles are near the longitudinal edge of the slab, midway between the transverse joints, a critical tensile bending stress occurs at the bottom of the slab under the wheel load. This stress increases greatly when there is a high positive temperature gradient through the slab (the top of the slab is warmer than the bottom of the slab). Repeated loadings of heavy axles under those conditions result in fatigue damage along the bottom edge of the slab, which eventually result in a transverse crack that propagates to the surface of the pavement. A reasonable standard error of the estimate for total transverse cracking or total percent slabs cracked is seven percent. The MEPDG predicts the total percent slabs cracked which includes both bottom-up and top-down cracking of JPCP.

Transverse Cracking, Top-Down (JPCP)—Repeated loading by heavy truck tractors with certain axle spacing when the pavement is exposed to high negative temperature gradients (the top of the slab cooler than the bottom of the slab) result in fatigue damage at the top of the slab. This stress eventually results in a transverse or diagonal crack that is initiated on the surface of the pavement. The critical wheel loading condition for top-down cracking involves a combination of axles that loads the opposite ends of a slab simultaneously. In the presence of a high negative temperature gradient, such load combinations cause a high-tensile stress at the top of the slab near the critical pavement edge. This type of loading is most often produced by the combination of steering and drive axles of truck tractors and other vehicles with similar axle spacing. Multiple trailers with relatively short trailer-to-trailer axle spacing are the other source of critical loadings for top-down cracking.

3.0 Significance and Use

Predicting pavement distress is a very complex process that involves uncertainty, variability, and approximations of all input parameters. Mechanistic concepts do provide a more rational and realistic methodology for accounting for variations and approximations, but all prediction models have errors associated with them. The overall error is termed the standard error of the estimate (s_e).

The goal of any calibration-validation process is to confirm that the prediction model can predict, without bias, pavement distress and smoothness, and to determine the standard error associated with the prediction equations. The standard error is used to establish confidence intervals for the prediction model which is used in the design reliability procedure. The standard error estimates the scatter of the data around the line of equality between the predicted and observed values of distress.

All prediction models in the MEPDG were globally calibrated using a representative sample of pavement test sites around North America. Most of these test sites are included in the LTPP program and were used because of the consistency in the monitored data over time and the diversity of test sections spread throughout North America. Policies on pavement preservation and maintenance, construction and material specifications, and materials, however, vary across the United States and can significantly affect distress and performance. These factors are not considered directly in the MEDPG, but can be considered indirectly through the local calibration parameters included in the MEPDG determined through local calibration.

This guide can be used to determine if local policies and practices result in a significant bias in the predicted values, and to recalibrate the MEPDG to local conditions and materials that were not considered in the global calibration process. Eliminating any significant bias and decreasing the standard error of the estimate will reduce construction costs at the same reliability level.

The local calibration process only relates to the transfer functions or statistical models (refer to Figure 1-1). The supporting, mathematical models within the MEPDG simulation model are assumed to be accurate and a correct simulation of real-world conditions. These supporting models are used to compute specific parameters that are needed to predict pavement distress, and include the Integrated Climatic Model (ICM), the structural response model (JULEA for flexible pavements and ISLAB200 for rigid pavements), and time-dependent material property models (strength-gain model of PCC and the age-hardening model for HMA). The time-dependent material property models are statistical models, but were assumed to be mathematical models in the global calibration process. Any error resulting from inaccuracies in the supporting statistical models, however, will be translated into lack-of-fit or model errors of the transfer functions.

4.0 Defining Accuracy of MEPDG Prediction Models

This section provides an overview of the general calibration and validation process of mechanistic-based simulation models for pavement design.

4.1 Calibration

The primary objective of model calibration is to reduce bias. A biased model will consistently produce either over-designed or under-designed pavements, both of which have important cost consequences. The secondary objective of calibration is to increase precision of the model predictions. A model that lacks precision is undesirable because it leads to inconsistency in design effectiveness, including some premature failures. As part of the calibration process, predicted distress is compared against measured distress and appropriate calibration adjustment factors are applied to eliminate significant bias and maximize precision in the model predictions.

In model calibration, a fitting process produces model constants that are evaluated based on goodness-of-fit criteria to decide on the best set of values for the coefficients of the statistical model formulated. The methods of evaluation are either: 1) an analytical process for models that suggest a linear relationship, or 2) the use of numerical optimization for models that suggest a non-linear relationship. The analytical calibration is based on the method of least squares using multiple regression analysis, stepwise regression analysis, principal components analysis, and/or principal component regression analysis. *NCHRP Results Digest #283* provides limited discussion on calibration and the use of different analytical and statistical techniques to reduce bias and determine the standard error for a particular transfer function (NCHRP, 2003).

Numerical optimization, consisting of unconstrained minimization techniques, can be defined as rank ordered or positive pattern search methods. Rank ordered or positive pattern search methods are not equivalent, but both will generally find an optimal set of calibration parameters for a specific set of site conditions and design features. Rank ordered based search methods consider the potential for the interrelationship between different calibration parameters of the transfer function. Pattern search methods are more commonly used, and rely on the steep descent or ascent procedure of nonlinear minimization and are gradient related. The step length or size of the gradients is important in the pattern search techniques to find the global minimum, rather than a localized minimum of mulitple calibration parameters. Numerical optimization methods, however, require a larger number of sections or observations and many more runs of the MEPDG to deterimine the set of calibration parameters that result in the lowest global error of each transfer function. For example, numerical optimization can require more than four times the number of runs needed for the analytical process.

Use of the analytical process within a constrained area or set of boundary conditions should provide reasonable results for the MEPDG transfer functions considering the measurement errors for each of the distresses and performance indicators predicted by the MEPDG. Measurement error and other components of the standard error of the estimate term are discussed in the next section of this document.

Two different calibration approaches may be required depending upon the nature of the distress being predicted through the transfer function. One approach is used for those models that directly calculate the magnitude of the surface distress, while the other approach is used for those models that calculate the incremental damage index rather than the actual distress magnitude. Both are briefly defined below.

Computation of Actual Distress Magnitude from Pavement Response. The term calibration refers to the mathematical process by which the difference between an observed result (e.g., the measured mean rut depth in a pavement section) and a predicted result (e.g., predicted mean rut depth from a permanent deformation model) is reduced to a minimum value for all available sections. The pavement response parameter is used to compute the incremental distress in a direct relationship. This fitting of the predicted to the observed results is most often accomplished by minimizing some function of the differences between the predicted and observed results (normally written as ε_i) by modifying the values of empirical parameters that are part of the model. These empirical parameters are necessary to compensate for the model simplification and limitations in simulating actual pavement behavior and distress development. This difference term $\varepsilon_i(b)$ can be defined by Eq. 4-1.

$$\varepsilon_i(b) = y_i - n(x_i; b) \qquad (4\text{-}1)$$

Where y_i is the ith observed response and $n(x_i; b)$ is the ith predicted response, the x_i are the independent variables that govern the predicted response, and b represents the calibration parameters or coefficients that are chosen such that the predicted responses are as close as possible to the observed responses.

For example, the MEPDG permanent deformation model directly predicts the magnitude of the actual pavement distress, the rutting measured at the pavement surface. The difference $\varepsilon_i(\theta)$ between the field rutting measurements and the model rutting predictions can be defined as $\varepsilon_i(\theta) = y_i - \theta$ $(x_i; \theta)$, where y_i is the ith observed response and η $(x_i; \theta)$ is the ith predicted response, the x_i are the independent variables that govern the predicted response, and θ represents the calibration parameters or coefficients that are chosen such that the predicted responses are as close as possible to the observed responses; i.e., that minimize $\varepsilon_i(\theta)$ in some overall sense.

Computation of Incremental Damage from Pavement Response. The incremental damage index is computed using a mathematical process describing the development of the distress in terms of accumulated damage. The pavement response is used to compute damage and damage is then correlated to the observed amount of distress. The field test sections were used to adjust or relate the cumulative incremental damage computations to the actual distress measured along the test sections at different points in time. Thus, the calibration proceeds by regressing the damage indices to the actual observations of distress. This approach determines one calibration or transfer function, which can include various site and pavement design features.

For example, the MEPDG model for fatigue cracking is based on an incremental damage index rather than the actual distress magnitude; i.e., the area of cracking. In this case, the empirical calibration coefficients attempt to relate measured cracked pavement area (the actual field distress) to the cumulative damage values (i.e., the model predictions).

Data collected from field test sections are used with both approaches to establish calibration coefficients such that the standard error is minimized between the predicted response (n) and the observed response (y).

4.2 Validation

The objective of model validation is to demonstrate that the calibrated model can produce robust and accurate predictions of pavement distress for cases other than those used for model calibration. Validation typically requires an additional and independent set of in-service pavement performance data. Successful model validation requires that the bias and precision statistics of the model when applied to the validation data set are similar to those obtained from model calibration. The purpose of validation is to determine whether the calibrated conceptual model is a reasonable representation of the real-world system, and if the desired accuracy or correspondence exists between the model and the real-world system.

The success of the validation process can be gauged based on the biases in predicted values and the standard error of estimate, s_e. The s_e for the validation may not be equal to the s_e for calibration; generally, it is higher. To test if it is significantly higher at a given level of significance, which would suggest that the validation failed, a chi-square test is typically used. Conversely, an operational definition of "reasonable correlation" is that the null hypothesis (of equality) is accepted when the paired t-test is used to compare the observed and predicted responses at a confidence interval of 95 percent ($\alpha = 0.05$).

The validation factorials should be designed to statistically test the null hypothesis for each performance indicator. The null hypothesis is that the predicted distress is not statistically different from the actual measured distress. If the null hypothesis is true, then the error is determined using all data for that distress type (both the calibration and validation data sets). If the validation process results in the rejection of the null hypothesis at the chosen significance level, the soundness and completeness of the conceptual and the operational models must be re-evaluated. Further changes in the models require another round of calibration and validation to assure that the revised models are sufficiently accurate.

The benefit of a stringent, independent test on the accuracy of the calibrated model far outweighs the increased costs associated with obtaining two independent data sets. Thus, the split sample approach is typically used in the calibration and validation of statistical and simulation models. A typical split of a sample is 80/20 with 80 percent of the data used in calibration and 20 percent used for verification (chosen randomly of course).

4.3 General Approach to Local Calibration-Validation

Two approaches can be used to improve on the accuracy of the prediction model, including the MEPDG global calibration coefficients for local conditions, policies, and materials. These two approaches are termed the split-sample approach, which is the traditional approach used and discussed in Subsection 4.3.1, and an alternative procedure called jack-knifing.

4.3.1 Traditional Approach—Split-Sample

Some type of organized subdivision of pavement conditions is usually employed for model development, calibration, and validation, because of the wide range of materials (e.g., natural subgrades, local aggregates), truck traffic levels, and environments (e.g., temperature ranges, rainfall levels) for which pavements must be designed. Typically, each cell within an experimental matrix would contain several field sites for which in-service pavement performance data exist for use in model calibration and validation. Alternatively, the same underlying performance model can be used for all cells with each cell being calibrated separately using a portion of the field sites for the cell, with the remaining field sites reserved for subsequent model validation.

Unfortunately, the most common procedure for model development is to use all of the data for calibrating the coefficients and to then take the resulting goodness-of-fit statistics (e.g., the correlation coefficient) as indicators of the prediction accuracy of the model. The calibration statistics consequently only reflect the accuracy of the model for regenerating the calibration data and may not accurately reflect prediction accuracy over the full population. This procedure ignores proper model validation and may produce misleading results unless the size of the calibration data set is exceedingly large, which is rarely the case for pavement performance data.

Those who recognize that calibration goodness-of-fit may not be a good indicator of prediction accuracy have often used split-sample testing for model validation. In the traditional application of split-sample testing, a portion of the data (typically half or more) is used for calibrating the coefficients while the remainder is used to validate accuracy. While split-sample calibration and validation is an improvement over no validation, it has some of the same limitations and can produce a misleading indication of model accuracy for small sample sizes (refer to Note 1). For small sample sizes, the jack-knifing approach is recommended as an alternate to the traditional split-sample approach.

Note 1—A small sample size is a relative term that depends on the number of factors included in the sampling template. In summary, a small sample size is defined as a partial factorial with less than 25 percent of the cells filled with a project but without replication.

4.3.2 Jack-Knife Testing—An Experimental Approach to Refine Model Validation

Jack-knifing is a procedure that provides more reliable assessments of model prediction accuracy than either traditional split-sample validation or the use of the calibration goodness-of-fit statistics. Jack-knifing provides goodness-of-fit statistics that are based on predictions, unlike the calibration statistics that depend on the data used for fitting the model parameters. Thus, the model validation statistics are developed independently of the data used for calibration. Multiple jack-knifing is used to assess the sensitivity of the validation goodness-of-fit statistics to sample size.

To develop jack-knife statistics from a sample of n sets of measured values, the data matrix is divided into two groups, one part for calibration and the other for prediction. These sets are selected at random. Assume that the data matrix includes measurements of p predictor variables X_{ij}, $j=1...p$ and a single criterion variable Y_i, with $i=1...n$ sets of measured values. Thus, the data matrix has n rows and $p+1$ columns.

For an n-1 jack-knife validation, the procedure begins by removing one set of measurements from the data matrix and calibrating the model with the remaining n-1 sets of measurements. The kth set of measurements that was withheld is then used to predict the criterion variable Y_k from which the standard error (e_1) is computed as the difference between the predicted (Y_k) and measured (Y_k) values of the criterion variable. A second set of measurements is removed while replacing the first set, and the new n-1 set is used to calibrate a new model. This new calibrated model is then used with the withheld set of X values to predict Y and compute the standard error, e_2.

The process of withholding, calibrating, and predicting is repeated until all n sets have been used for prediction. This yields n values of the standard error, from which the jack-knifing goodness-of-fit statistics can be computed. While both the calibration statistics based on all n sets and the jack-knifing statistics are computed from n measures of the error, the jack-knifed errors are computed from measured X values that were not used in calibrating the model coefficients. Thus, the jack-knifing goodness-of-fit statistics are considered to be independent measures of model accuracy.

Because sample sizes of most pavement engineering data sets are limited, one objective of model validation is to assess the sensitivity, or stability, of the accuracy of the model to sample size. To assess the stability of the jack-knifed goodness-of-fit statistics, multiple jack-knifing can be performed by withholding two sets of X, while calibrating on the remaining n-2 sets. Two errors are computed for each calibration based on the n-2 withheld sets of X. For small samples, the goodness-of-fit statistics for the n-2 jack-knifing may be quite different from those for the n-1 jack-knifing. If however the n-1 and n-2 jack-knifing goodness-of-fit statistics are similar, this indicates that the n-1 jack-knifing statistics are not sensitive to the sample size and the statistics are stable. Stable statistics are reliable indicators of goodness-of-fit or prediction accuracy.

The primary advantage of jack-knifing is that the goodness-of-fit statistics are based on predictions from data that are independent of the calibration data. Thus, they more likely indicate the accuracy of future predictions than the statistics based on calibration of all n data vectors. The use of multiple jack-knifing to assess the stability of the prediction statistics is a second advantage of jack-knifing. A third advantage is that the method is easy to apply.

Split-sample validation differs from jack-knifing in that the goodness-of-fit statistics for both calibration and prediction are based on $n/2$ values (for symmetric split sampling the usual case) rather than n values. Traditional split-sample validation has the distinct disadvantage that, if n is small relative to the inference space being simulated, then $n/2$ is even smaller, which produces inaccurate calibrations, inaccurate coefficients, and less reliable prediction accuracy.

To overcome in part this deficiency, a method was proposed as part of the NCHRP Project 9-30 experimental plan that combines jack-knifing and split-sample testing (NCHRP, 2003b). It is essentially an *n*/2 jack-knifing scheme and will be termed split-sample jack-knifing. Split-sample jack-knifing provides somewhat better measures of prediction accuracy than the traditional split-sample validation. This approach and process is recommended for the local calibration-validation process because the sample size for local calibration will probably be much smaller than used for the global calibration process.

5.0 Components of the Standard Error of the Estimate

The standard error of the estimate of a prediction model is an important factor that must be understood and quantified in making a decision on whether to try and increase the precision of a simulation model. This section defines and describes the four major components of variance, which are listed below and in Table 5-1.

1. Measurement error.
2. Input error.
3. Model or lack-of-fit error.
4. Pure error.

These components of the total standard error can be mathematically expressed by the sum of the error variances, as shown by Eq. 5-1, assuming that no correlation exists between the contributing errors to the overall error.

$$(V_{Total})^2 = (V_m)^2 + (V_{Input})^2 + (V_{Pure})^2 \qquad (5\text{-}1)$$

where:

V_{Total} = Total variance of the residual error of prediction associated with the "actual" versus "predicted" performance quantity, sometimes referred to as the calibration error variance,

V_{Input} = Portion of the total variance caused by variations in laboratory and field measurements to estimate the model inputs,

V_m = Portion of the total variation caused by inaccuracies in measuring the distress along the test section used in the calibration process,

V_{Purel} = Portion of the total variance due to replication, referred to as pure error, and

V_l = Portion of the total variance caused by inadequate theory, algorithms, and/or an incorrect model form, typically referred to as lack-of-fit or model variance.

Table 5-1. Summary of Major Components of Calibration Error Dependency

Components of Calibration Error	Error Is:		
	Distress/IRI Dependent	**Input Level Dependent**	**Prediction Model Dependent**
Measurement Error	Yes	No	No
Input Error	No	Yes	No
Model Error	No	No	Yes
Pure Error	Yes	Yes	Yes

This section discusses the importance of the error terms resulting from the calibration of the MEPDG distress prediction equations or transfer functions. The separation and quantification of the sources of variability is important when refining the calibration-validation process to reduce the total standard error. As an example, decreasing the input error of pavement material properties will have little effect on reducing the total standard error or uncertainty of the predictions for the condition if the majority of the total error is caused by measurement error of the distress observations. Most likely, one would want to implement the calibration refinement that has the greatest effect on reducing the total error for the least cost. Each of the four error components is discussed in the following sections.

5.1 Distress/IRI Measurement Error

Errors associated with measuring the distress quantity and IRI for a pavement section are defined as measurement errors. For example, the mean rut depth measured along a project is not the true value, but an estimate of the true mean of the population or test section at a particular point in time. The greater the number of measurements within a test section, the lower the potential difference between the sample and population means and the lower the measurement error. Measurement errors are dependent on the performance measure being calibrated, but are independent of the input level and prediction equation (refer to Table 5-1). The variance in the measured value is composed of different parts, which are listed below.

- V_{mr}—The variance in measurement at a point determined by taking multiple readings at the same location. This component of the measurement variance decreases as the number of repeat measurements increase at the same point on the roadway.
- V_{ms}—The variance in taking a measurement at the same test point (location) but at different times, or the expected difference in taking readings at the same point of the pavement's surface.
- V_{mv}—The variance in taking measurements along the project or the inherent error—sample versus population mean. This component of the measurement error decreases as the number of point measurements increase along a roadway segment.

All of these variances are assumed to be independent and can be added together to determine the total variance in the measured value. These variances are also constant in developing and/or comparing different prediction models using the same database. In general, the sample and test components of the measurement variance are small relative to the inherent variance along the project. However, the measured error or variance for each distress must be representative of how the data were used in the calibration-validation process. For example, if smoothed and cleaned data were used, then the variance associated with the smoothed and cleaned data must be used with the calibration error or variance. The measured data should be used to determine the true bias and standard error of the estimate. The use of smoothed data is not recommended. The bias and precision components (repeatability error) of the distress measured within LTPP were evaluated and documented by Rada, et al. (Rada, 1999).

5.2 Estimated Input Error

The errors associated with estimating each input parameter needed to predict the performance indicator and used in the calibration process are defined as input errors. For example, the mean asphalt content by volume is a required HMA property for predicting mean rut depth or fatigue cracking over time. The mean asphalt content by volume for each HMA layer within an LTPP test section is determined by averaging no more than two test values for that HMA layer. The mean asphalt content of these two tests is not the true mean value for that layer, it is only an estimate based on the results from two tests. The input error is dependent on the material property (or input level, because the input level defines the material properties and parameters to be used) required to predict the performance measure, and is independent of distress type and prediction equation.

The input variance component is composed of three basic parts; testing error, sampling error, and inherent variation of the material along a project. All of these variance components are independent and additive to determine the total input error component. In general, the test and sample error parts of this component are small relative to the inherent material variance or error along the project or test section.

5.3 Model or Lack-of-Fit Error

The inability of a model to predict the actual or true value of the performance measure due to deficiencies in the transfer function or inappropriate assumptions included in the mathematical model, and its inability to model real world conditions, are defined as model or lack-of-fit errors. This type of error is a result of inappropriate assumptions, model simplicity, or inadequate damage algorithm including functional form for the predicted performance indicator. For example, assuming uniform contact pressure under the wheel loads on a flexible pavement is not reality—it is a simplification of reality. Lack-of-fit errors are dependent on the prediction and response (mathematical) models and are independent of the input level and performance measure.

5.4 Pure Error

The random or normal variation between distress values of supposedly "identical" roadway segments is defined as pure error (refer to Note 2). Pure error is dependent on the input level, distress type, and prediction equation. Pure error is difficult to quantify unless replicate test sections within each experimental cell are included in the calibration factorial, which was not the case for the MEPDG distress prediction models.

Note 2—Identical roadway segments are usually restricted to test sections built under highly controlled conditions—test tracks or test pads placed at accelerated pavement testing (APT) facilities. Identical as used in this context is a relative term, which implies that multiple projects or roadway segments with similar experimental factors fall within the same cell and have the same range of secondary properties or factors. Replicate roadway segments do not necessarily have to have the same properties to be considered identical.

6.0 Step-by-Step Procedure for Local Calibration

This section lists and defines the steps that are suggested for calibrating the MEPDG to local conditions, policies, and materials. These steps are shown in the form of a flow chart in Figures 6-1 and 6-2.

Step 1—Select Hierarchical Input Level for Each Input Parameter

The first step in the local calibration process is to select the hierarchical input level for the inputs that will be used by an agency for pavement design and analysis. This step will likely be a policy decision, influenced by the agency's current field and laboratory testing capabilities, material and construction specifications, and traffic data collection procedures and equipment. The *MEPDG Manual of Practice* provides recommendations on selecting the hierarchical input level for each input parameter (AASHTO, 2008).

Selecting the hierarchical input level can be important because decisions made in this step may have a significant impact on the final standard error of each distress prediction model, which affects material quality requirements and construction costs. If the input error only has a minor effect on the standard error of the estimate for the transfer function, the input level may only have a minor effect on the design and the construction costs. (Refer to discussion in Section 5, Components of the Standard Error of the Estimate.) The highest level of input data available from the LTPP database were used to determine the inputs for the global recalibration effort under NCHRP Project 1-40D (NCHRP, 2006), and resulting standard error of the estimate. Agencies will probably elect to use different input levels for some of the input parameters. The bias and standard error of the estimate should be determined for the input levels that will be typically used by an agency for pavement design.

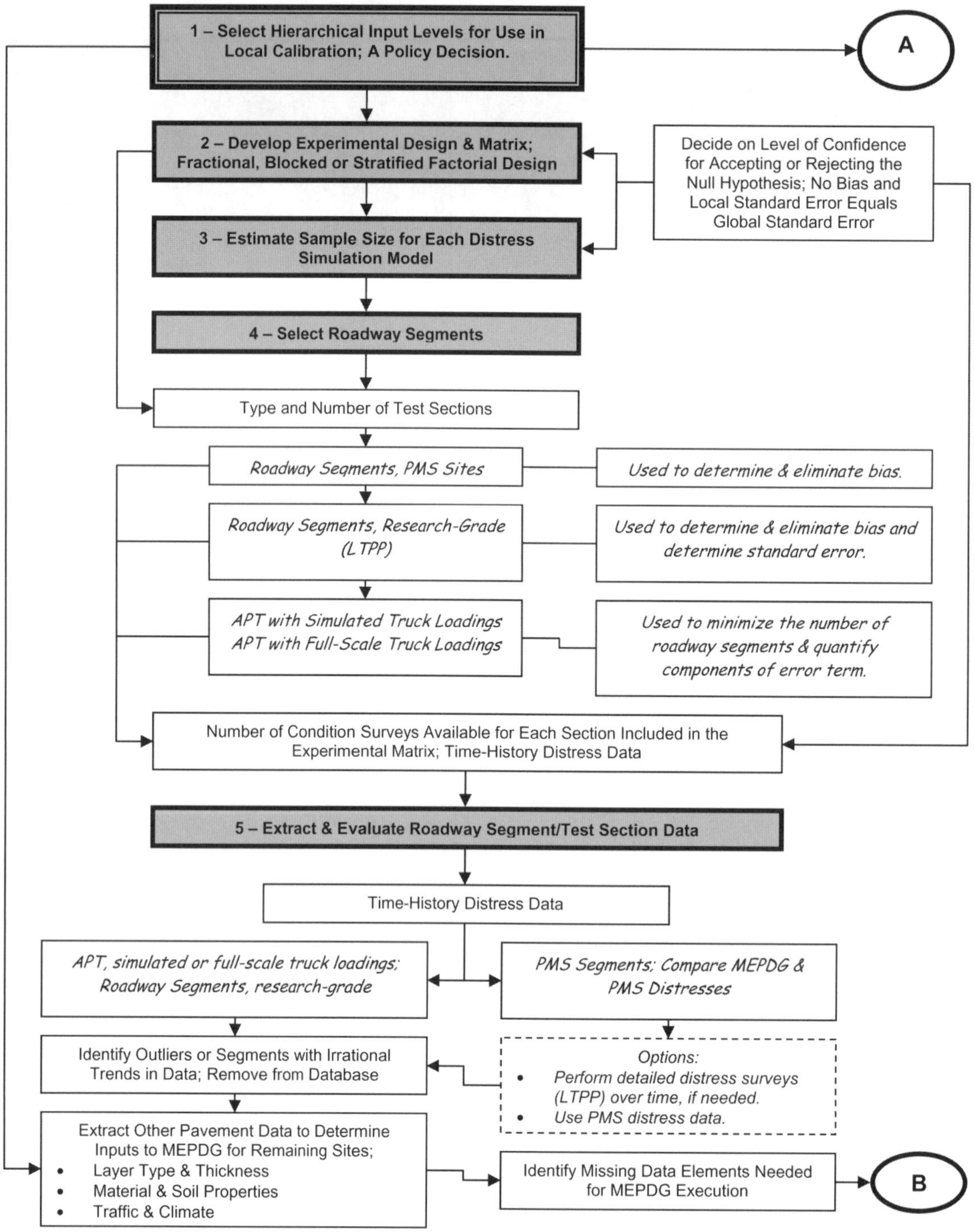

Figure 6-1. Flow Chart of the Procedure and Steps Suggested for Local Calibration; Steps 1 Through 5

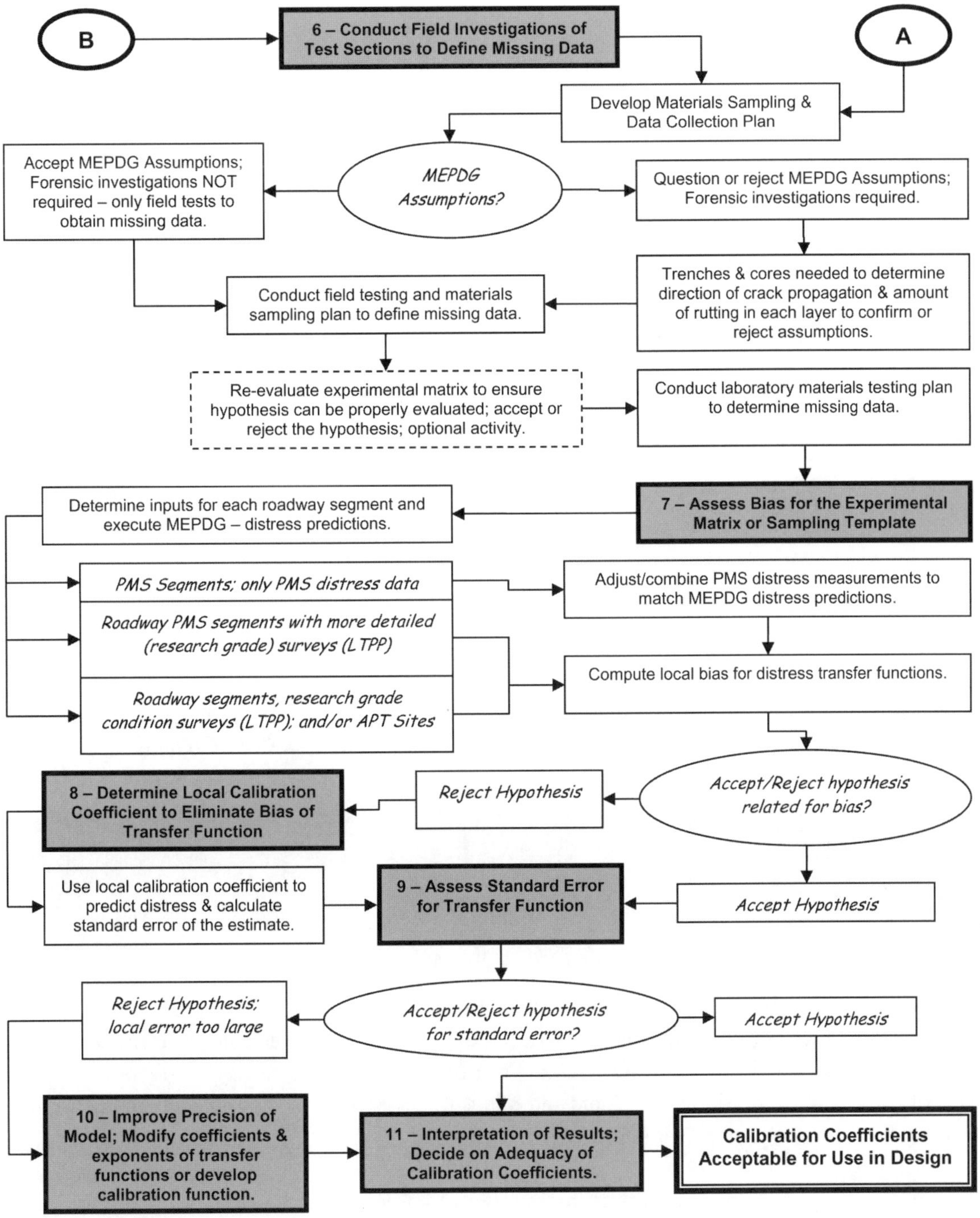

Figure 6-2. Flow Chart of the Procedure and Steps Suggested for Local Calibration; Steps 6 Through 11

Step 2—Develop Local Experimental Plan and Sampling Template

The second step is to develop a detailed, statistically sound experimental plan or sampling template to refine the calibration of the MEPDG distress and IRI prediction models based on local conditions, policies, and materials, if required. The local or regional calibration factorial for each distress simulation model should be designed to accomplish three objectives:

1. Determine whether there is any local bias in the MEPDG distress predictions or simulation model.
2. Establish the cause of any bias, if it is found through the local validation process.
3. Determine the local calibration coefficients or function for each distress and IRI prediction models.

The primary and secondary tiers within the sampling template should be based on the agency's standard practice and specifications for the more common new construction and rehabilitation strategies. The primary tier parameters in the sampling template should be distress dependent, and likely include pavement type, surface layer type and thickness, and subgrade soil type. Each column in the matrix should represent the effect of changing structure (layer thickness), mixtures, and foundation soils. The secondary tier parameters should include climate (temperature), traffic, and other design features that are pavement type dependent. These are considered secondary parameters because traffic is probably interrelated to surface thickness and climate is interrelated to asphalt binder grade. In some areas, climate may need to be included as a primary tier parameter because of large variations in climate within the same region (e.g., the western part of the United States that includes mountains and plains).

The sampling template should be designed as a fractional factorial matrix as much as possible. Not all cells will likely be filled with or without replicate roadway segments. The matrix should be a balanced design that can be blocked for specific design features or site conditions for each type of pavement and distress. Blocking the fractional factorial will determine whether the bias and standard error of the transfer function is dependent on any of the primary tier parameters of the matrix. This type of matrix is recommended because the experiment needs to evaluate the effect of pavement type and local conditions and materials on reducing the bias and standard error term. When this is completed, answer the question: "Is the bias and/or standard error of the transfer function dependent on type of soil, surface mixture properties, climate, etc.?"

Most cells within the sampling template or matrix should contain two replicate projects to provide an estimate of the pure error.

Step 3—Estimate Sample Size for Specific Distress Prediction Models

The numbers in each cell of the sampling template should include replicate projects, as noted in Step 2. This step is used to estimate the sample size or number of roadway segments to confirm the adequacy of the global calibration coefficients and determine the local calibration coefficients for a specific distress prediction model, if needed.

Both bias and precision are important, thus the number of model evaluations (i.e., the sample size) needed to properly validate the prediction model is evaluated for both bias and precision. The bias is the average residual error; therefore, the confidence interval on the mean can be used to relate the sample size and the bias. Letting e_t be the tolerable bias, the confidence interval on the mean yields the following expression:

$$N \geq \left(\frac{ts_y}{e_t}\right)^2 \qquad (6\text{-}1)$$

Where s_y is the standard deviation of the "true" values of Y and t is based on n-1 degrees of freedom. For accuracy, the standard error of estimate will be used. Since the square of s_e is a variance, the confidence interval on the variance can be used to show the relationship between sample size and the relative error variance (s_e/s_y). The basic equation for the confidence interval is:

$$\sigma^2 \geq (n-1)\frac{s^2}{x_a^2} \qquad (6\text{-}2)$$

Where the chi-square statistic is n-1 degrees of freedom and level of significance, α. Inserting s_e and s_y yields:

$$\frac{s_e}{s_y} \geq \left[\frac{x_\alpha^2}{n-1}\right]^{0.5} \qquad (6\text{-}3)$$

By selecting a level of significance, the relative deviation s_e/s_y can be determined for a selected sample size. Three levels of significance can be used in estimating the sample size for each distress: 75, 90, and 95 percent. A level of significance of 90 percent is suggested as a practical level in determining the sample size to be used in the experiment.

The same test sections can and should be used for all distresses, because of the coupling effect between different distresses. The MEPDG assumes that all distress transfer functions are uncoupled (distress occurrence and magnitude is independent of the other distresses), with the exception of the IRI regression equation or statistical relationships. The following provides guidance for the minimum number of total test sections for each distress.

- Distortion (Total Rutting or Faulting)—20 roadway segments
- Load-Related Cracking—30 roadway segments
- Non-Load-Related Cracking—26 roadway segments
- Reflection Cracking (HMA surfaces only)—26 roadway segments

The bias and precision components of the distresses reported within the LTPP program were used in estimating the number of test sections listed above (Rada, 1999). If repeatedability errors are unavailable to the agency or user, those reported within the LTPP program for each distress may be used.

Step 4—Select Roadway Segments
This step is used to select roadway projects to obtain maximum benefit of existing information and data to keep sampling and field testing costs to a minimum.

As noted above, replicate projects should be included in specific cells of the sampling template for a specific distress. One of the replicate test sections can come from one of the other distress matrix or factorials. Test sections for refining the validation process will be used for multiple purposes or distresses for efficiency. For example, the test sections that exhibit relatively high amounts of fatigue cracking can be used in the rutting matrix or thermal cracking matrix for low magnitudes of those distresses.

Although three types of experimental test sections can be used in the local validation-calibration refinement plan, roadway segments or long-term field experiments need to be used to determine the standard error of the estimate for each distress simulation model. The three types of experimental test sections are listed and defined below.

1. Long-term, full-scale roadway segments or test sections should be used to fully validate and calibrate the distress prediction models and confirm the superposition of the environmental, aging, and wheel-load effects on the predictions of distress. All of the test sections included in the NCHRP Project 1-40D calibration efforts were from this category of pavement test sites (NCHRP, 2006). Long-term roadway segments can be grouped into two types—those that are PMS segments and those that are research-grade roadway segments (e.g., LTPP sites). Although most of the roadway segments included in the NCHRP Project 1-40D were LTPP test sections (research grade sites), it is expected that many agencies will use PMS segments for judging the adequacy of the global calibration coefficient to their local policies, conditions, and materials. Both types of roadway segments are discussed further in Step 7, and are recommended for use in determining the standard error of the estimate for all distress prediction models.

2. Accelerated Pavement Testing (APT) pads with simulated truck loadings can be used for rapid verification of the form of the distress growth models (transfer function) and selected factor effects on the occurrence of distress. Results from APT test pads are independent of climatic-related factors and time-dependent properties of the pavement materials, so fewer tests are needed to determine the effect of selected factors. APT sites can be used to supplement the roadway segments used in the local calibration process, but should not be used to determine the standard error of the estimate. Use of APT test pads will result in much lower standard errors of the estimate, because traffic and climate parameters are highly controlled and time-dependent properties are excluded from these short-term loading conditions. Thus, APT test pads should only be used to determine bias and to quantify the variance components of the distress prediction model.

3. APT experiments with full-scale truck loadings (test tracks) can be used to calibrate and validate the effects of wheel load on the distress predictions without the added complexity of long-term aging and extensive environmental variations. Results from this type of experiment are slightly dependent on the climatic factors and time-dependent material properties. Use of these full-scale APT experiments will also result in lower standard errors of the estimate, because many of the factors are controlled. Results from full-scale APT experiments should be used to supplement and reduce the number of roadway segments required for local calibration in determining bias and to quantify the error term components of the distress prediction mode.

A listing of some factors that should be considered in selecting roadway segments for use in the local validation-calibration refinement plan includes the following:

- Roadway segments should be selected with the fewest number of structural layers and materials (e.g., one PCC layer, one or two HMA layers, one unbound base layer, and one subbase layer) to reduce the amount of testing and input required for material characterizations. These roadway segments, however, need to include the types of new construction and rehabilitation strategies typically used or specified by the agency. The roadway segments used to define the standard error of the estimate should include the range of materials and soils that are common to an area or region and the physical condition of those materials and soils.

- Roadway segments with and without overlays are needed for the validation-calibration sampling template. Those segments that have detailed time-history distress data prior to and after rehabilitation should be given a higher priority for use in the experiment because these segments can serve in dual roles as both new construction and rehabilitated pavements.

- Roadway segments that include non-conventional mixtures or layers should be included in the experimental plan to ensure that the model forms and calibration factors are representative of these mixtures. Non-conventional mixtures can include: stone matrix asphalt (SMA), polymer modified asphalt (PMA), open-graded drainage layers, cement-aggregate mixtures, and high-strength PCC mixtures. Many of the LTPP test sections included in the NCHRP Project 1-37A calibration factorial were built with conventional HMA and PCC mixtures. The flexible sections excluded open-graded drainage, SMA, and PMA layers. There were numerous sections with open-graded mixtures in the JPCP sections. The *MEPDG Manual of Practice* provides a more completed listing on the limitations of the MEPDG and those design features not considered directly by the MEPDG (AASHTO, 2008).

It is recommended that at least three condition surveys be available for each roadway segment to estimate the incremental increase in distress over time. The interval between the distress measurements should be similar between all of the test sections. It is also suggested that this time-history distress data represent at least a 10-year period, if available. This time period will ensure that all time-dependent material properties and the occurrence of distress are properly taken into account in the determination of any bias and the standard error of the estimate. If available, repeat condition surveys should be planned for those roadway segments that exhibit higher levels of distress to reduce the inherent variability of distress measurements and estimate the measurement error for a particular distress. A similar number of observations per age, per project should be considered in selecting roadway segments for the sampling template.

Step 5—Extract and Evaluate Distress and Project Data

This step of the local calibration process is to collect all data and identify any missing data elements that are needed to execute the MEPDG. All data should be entered into a calibration database, similar to the one that was developed under NCHRP Project 9-30 for flexible pavements (NCHRP, 2003a), or at least filed for future reference. This step is grouped into four activities, as discussed in the following paragraphs.

Step 5.1

The first activity under Step 5 is to extract, review, and convert the measured distress data into the values predicted by the MEPDG, if needed. It is imperative that a consistent definition and measurement protocol of surface distress be used throughout any calibration-validation process. If possible, all flexible pavement distress data should be measured in accordance with AASHTO R 55 *Standard Practice for Quantifying Cracks in Asphalt Pavement Surface* and AASHTO R 48 *Standard Practice for Determining Rut Depth in Pavements* or the FHWA LTPP publication *Data Collection Guide for Long-Term Pavement Performance.* All rigid pavement distress data should be measured in accordance with the FHWA LTPP publication *Data Collection Guide for Long-Term Pavement Performance* and the AASHTO R 36 *Standard Practice for Evaluating Faulting of Concrete Pavements.* Distress measurements should be made to ensure consistency with the MEPDG predictions of distress and smootheness (FHWA, 2003). Pavement smoothness should be measured in accordance with AASHTO R 57 *Standard Practice for Operating Inertial Profilers and Evaluating Pavement Profiles.*

Many agencies, however, will need to use their PMS distress data for the local validation-calibration effort, which may differ from LTPP. For this condition, two options are available for use by the agency.

1. The first is to select PMS segments (refer to Step 4) and complete distress surveys in accordance with the *LTPP Distress Identification Manual* (FHWA, 2003). This option is time consuming to collect sufficient distress data for completing the local calibration. Agencies that select this option will need to have at least a five-year implementation plan in place for the MEPDG to ensure a minimum of three observations per project. Few agencies are expected to select this option.

   Distress surveys can be completed on the selected PMS segments within one season to reduce the time, but time-history distress data will be unavailable for any one PMS segment. This single-distress point is not suggested for use, because the incremental increase in distress over time will not be included in the evaluation of bias and in determining of standard error of the estimate for the distress prediction models. The lack-of-fit error will not be well-defined for this option.

2. The second, and probably preferred, option is to select PMS segments and use the PMS condition survey data in the local validation-calibration effort. This option requires less time and cost, but the PMS measurements may need to be adjusted or modified to be consistent with the MEPDG distress predictions (refer to Step 7).

Step 5.2

The second activity under this step is to compare the maximum measured distress values to the trigger values or design criteria used by the agency for each distress. The average maximum distress values from the sampling template should exceed 50 percent of the design criteria, as a minimum. This consideration becomes important when evaluating the bias and standard error terms of the prediction model under Steps 7 and 9, respectively. If the maximum distress values are significantly lower than the agency's design criteria for that distress (less than 50 percent of the design criteria), the accuracy and bias of the transfer function may not be well defined at the values that trigger major rehabilitation.

Step 5.3

The measured distress data for all roadway segments should be evaluated and checked for anomalies and outliers—observations that have irrational trends in the distress data. This evaluation can be limited to visual inspection of the data over time to ensure that the distress data are reasonable, or include a detailed statistical comparison of the performance data. Multiple distress surveys and profile measurements are used to establish the performance trends for each roadway segment. Any segment with irrational trends in the distress data should be considered for removal from the local calibration database. As a minimum, the following two questions should be asked in evaluating the measured distress data.

1. Does the data make sense within and between each roadway segment? Obviously, any zeros that represent non-entry values should be removed from the local validation-calibration database. Distress data that return to zero values within the measurement period may indicate some type of maintenance or rehabilitation activity.

- Measurements taken after structural rehabilitation should be removed from the database or the observation period should end prior to the rehabilitation activity.
- Distress values that are zero as a result of some maintenance or pavement preservation activity, which is a part of the agency's management policy, should be removed but future distress observation values after that activity should be used.

2. Are there roadway segments with anomalies, outliers, or blunders in the data? If the outliers or anomalies can be explained and are a result of some non-typical condition, they should be removed. If the outliers or anomalies cannot be explained, they should remain in the database.

For the roadway segments that remain, all data should be extracted for use in determining the required inputs for the hierarchical input levels selected (refer to Step 1). Data sources that will likely be used by at least some agencies to determine the MEPDG inputs are construction records, acceptance tests in a quality assurance (QA) program, and as-built construction plans. Use of QA and historical data provides overall project or lot averages that can be different from the layer properties of individual PMS segments. The difference between PMS segments and lot average values will increase the input error component of the total standard error term (see Section 5).

Each agency will need to consider these sources of errors and make a judgment decision on whether to increase the effort and costs in determining the inputs to the MEPDG for local calibration. The final standard error of the estimate for each distress simulation model will impact this judgment decision in the long-term because of its effect on construction costs.

Any missing or questionable data to determine the MEPDG inputs should be identified. The missing or questionable data elements should be determined through field investigations. For the PMS segments selected, the falling weight deflectometer (FWD) deflection basin and other field tests should be performed to confirm layer thickness and estimate the in-place modulus values for each structural layer. FWD testing should be performed by AASHTO T 256 *Standard Method of Test for Pavement Deflection Measurements.* The *MEPDG Manual of Practice* includes recommends for a field test program for pavement evaluation and rehabilitation studies.

Step 6—Conduct Field and Forensic Investigations

Step 6.1
The first activity of this step is to develop a materials sampling and testing plan to determine any missing data element or to validate some key inputs for the roadway segments selected. The *MEPDG Manual of Practice* provides recommended guidelines for field investigations.

The pavement materials should be recovered and tested in accordance with the agency's standard practice that is used during pavement evaluation for rehabilitation design. The *MEPDG Manual of Practice* does provide recommendations for both field and laboratory testing for measuring the layer properties in accordance with the hierarchical input level selected. If the agency's standards differ from the *MEPDG Manual of Practice*, the agency's standards should be followed because those are the ones that will be used for day-to-day new pavement and rehabilitation designs.

Step 6.2
As part of this step and any field investigation, an agency needs to decide whether forensic investigations are required to confirm the assumptions embedded in the MEPDG. As an example, the portion of total rutting measured at the surface that can be assigned to each pavement layer and the location of where cracks initiated (top-down versus bottom-up, load-related cracking). Two options are available that have a significant impact on costs and time to conduct any field investigation.

If the agency elects to accept the MEPDG assumptions for layer rutting and the location of crack initiation, no forensic investigations are required. For this option, the agency should restrict the local calibration to total rut depth and total load related cracking—combining longitudinal and alligator cracks within the wheel path (refer to Section 2.4).

If the agency rejects or questions the MEPDG assumptions under the first option, then trenches and cores will be needed to measure the rut depths within each pavement layer and estimate the direction of crack propagation. This option will likely require additional roadway segments and/or APT sections for confirming or adjusting the local calibration values for rutting and fatigue cracking. It should be noted that no trenches and cores were taken under NCHRP Projects 1-37A and 1-40D to verify and confirm the amount of rutting in each pavement layer, as well as where the cracks initiated or the direction of crack propagation.

Trenches or test pits are recommended so that individual pavement layer rutting can be measured, but are only needed for projects that have exhibited levels of rutting greater than 0.35 in. at the surface. It is difficult at best to measure the permanent deformation in subsurface layers for rut depths shallower than 0.35 in.

To determine the percentage of load-related cracks that start at the top of the pavement and propagate downward (as opposed to the classical assumption of bottom-up cracking), 6-in. diameter cores can be drilled directly on top of load-related cracks and extracted to observe the depth of the crack. Crack initiation and propagation direction should be reported for each core. Crack width at the initiation point and depth of crack from the initiation point should also be reported. If the crack extends completely through the HMA layers and it is impossible to determine the direction of crack propagation or where the cracks initiated, it is recommended that the agency assume that the cracks initiated at the bottom of the HMA layers.

Step 6.3
Prior to going to Step 7, the number of roadway segments remaining with all data needed to execute the MEPDG should be re-evaluated to ensure that a sufficient number of segments are available for the local validation-calibration effort. If too many of the roadway segments have been removed for one reason or the other, additional segments may need to be added to the sampling template.

Step 7—Assess Local Bias: Validation of Global Calibration Values to Local Conditions, Policies, and Materials
The MEPDG and global calibration values should be used to calculate the performance indicators for each roadway segment (new pavement and rehabilitation strategies). The predicted values are compared to the measured values to determine bias and the standard error of the estimate to validate each distress prediction model for local conditions, policies, specifications, and materials. The distresses predicted by the MEPDG for calibration puposes should be based on average values for

each input parameter. In assessing and eliminating the bias, if needed, the predicted distresses at a 50 percent reliability level should always be used. In other words, the average input values and the distresses at a 50 percent reliability level should be used within this step, as well as within Steps 8, 9, and 10.

Step 7.1
The bias and standard error of the estimate should be determined for this full set of data for each distress simulation model. Compare the predictions for each performance indicator to the measurements (or adjusted observations; refer to Step 5), and compute the residual errors, bias, and standard error of the estimate for each distress prediction model. A plot of the predicted values and measured data should be prepared to compare the general location of the data points to the line of equality.

Step 7.2
Evaluate the null hypothesis for the sampling template or experimental factorial (refer to Steps 2 and 3). The null hypothesis for this initial assessment is that there is no bias or no systematic difference between the measured and predicted values of distress. The null hypothesis should be evaluated for the entire sampling template and individual blocks within the sampling template. A paired *t*-test can be used to determine if there is a significant difference between the sets of predicted and measured distress and IRI values. The null hypothesis is as follows:

$$H_0 : \sum\left(y_{Measured} - x_{Predicted}\right) = 0 \tag{6-4}$$

where:
$y_{Measured}$ = Measured value, and
$x_{Predicted}$ = Predicted value using the model.

It is possible that the above hypothesis could be accepted (the sum of the residual errors are indifferent to zero), but the model still be biased. Two other model parameters (termed intercept and slope estimators) should be used to fully evaluate model bias using the following fitted regression model between the measured (*y*) and predicted (*x*) values, as well as the variability in the measured value associated with the distributed errors for each predicted value.

$$\hat{y}_i = b_o + m\left(x_i\right) \tag{6-5}$$

This regression model is used to provide estimators of the mean measured values ($\hat{y}_i$). The intercept (b_o) and slope (m) are used in hypothesis testing as follows:

H_o: $b_o = 0$
H_o: $m = 1.0$

In summary:

- If any of the null hypothesis is rejected, the specific distress prediction model should be recalibrated to the local conditions and materials—proceed to Step 8. The results from the three hypothesis tests can be used to make decisions during the re-calibration process.

- If the null hypothesis is accepted (no bias), the standard error of the estimate for the local data set should be compared to the global calibration data set—proceed to Step 9. The global standard errors are provided in the "Tools" section of the MEPDG software for each distress and in the *MEPDG Manual of Practice.*

Step 8—Eliminate Local Bias of Distress and IRI Prediction Models

The process used to eliminate the bias found to be significant from using the global calibration values depends on the cause of the bias and accuracy desired by the agency. In general, there are three possibilities, which are listed below.

1. The residual errors are, for the most part, always positive or negative with a low standard error of the estimate in comparison to the trigger value, and the slope of the residual errors versus predicted values is relatively constant and close to zero. The precision of the prediction model is reasonable but the accuracy is poor (large bias). In this case, the local calibration coefficient is used to reduce the bias. This condition generally requires the least level of effort and the fewest number of runs or iterations of the MEPDG to reduce the bias.

2. The bias is low and relatively constant with time or number of loading cycles, but the residual errors have a wide dispersion varying from positive to negative values. The accuracy of the prediction model is reasonable, but the precision is poor. In this case, the coefficient of the prediction equation is used to reduce the bias but the value of the local calibration coefficient is probably dependent on some site feature, material property, and/or design feature included in the sampling template. This condition generally requires more runs and a higher level of effort to reduce the bias.

3. The residual errors versus the predicted values exhibit a significant and variable slope that appears to be dependent on the predicted value. The precision of the prediction model is poor and the accuracy is time or number of loading cycles dependent—there is poor correlation between the predicted and measured values. This condition is the most difficult to evaluate because the exponent of the number of loading cycles needs to be considered. This condition also requires the highest level of effort and many more runs to reduce the bias.

The agency needs to first decide on whether to use the agency specific values or the local calibration parameters that are considered as inputs in the MEPDG software. Either one can be used with success. The following provides general guidance.

Compute the bias within each block of the sampling template (refer to Section 5) to determine whether the local bias is dependent on any primary or secondary tier parameter of the sampling template. Results from this analysis of local bias can be used to make revisions to specific calibration parameters to eliminate the local bias.

Adjust the local calibration values (agency specific values) for the distress transfer functions to eliminate the bias. Tables 6-1 and 6-2 list the local calibration parameters of the MEPDG transfer functions or distress and IRI prediction models that should be considered for revising the predictions to eliminate bias for flexible and rigid pavements, respectively. These tables are provided for guidance only in eliminating any local bias in the predictions. In addition, the local calibration values could be dependent on site factors, layer parameters, or policies established by the agency. If the local calibration values (agency specific values) are found to be dependent on some site factor, design feature, or material property, those types of adjustments or corrections need to be made external to the MEPDG.

Table 6-1. Recommendation for the Flexible Pavement Transfer Function Calibration Parameters to Be Adjusted for Eliminating Bias and Reducing the Standard Error

Distress		Eliminate Bias	Reduce Standard Error
Total Rutting	Unbound Materials and HMA Layers	k_{r1}, β_{s1}, or β_{r1}	kr_2, kr_3, and β_{r2}, β_{r3}
Load-Related Cracking	Alligator Cracking	C_2 or k_{f1}	k_{f2}, k_{f3}, and C_1
	Longitudinal Cracking	C_2 or k_{f1}	k_{f2}, k_{f3}, and C_1
	Semi-Rigid Pavements	C_2 or β_{c1}	C_1, C_2, C_4
Non-Load-Related Cracking	Transverse Cracking	β_{t3}	β_{t3}
IRI		C_4	C_1, C_2, C_3

Table 6-2. Recommendation for the Rigid Pavement Transfer Function Calibration Coefficients to Be Adjusted for Eliminating Bias and Reducing the Standard Error

Distress		Eliminate Bias	Reduce Standard Error
Faulting		C_1	C_1
Fatigue Cracking		C_1 or C_4	C_2, C_5
CRCP Punchouts	Fatigue	C_1	C_2
	Punchouts	C_3	C_4, C_5
	Crack Widths	C_6	C_6
IRI	JPCP	C_4	C_1
	CRPC	C_4	C_1, C_2

After the bias has been eliminated, compute the standard error of the estimate using the local calibration values (or agency specific values) based on local conditions; compare that standard error to the global standard error reported under NCHRP Project 1-40D and included in Section 5 of the *MEPDG Manual of Practice* (AASHTO, 2008)—proceed to Step 9.

Step 9—Assess the Standard Error of the Estimate

Compare the standard error determined from the sampling template to the standard error derived from the global data set, which are included in Section 5 of the *MEPDG Manual of Practice* for each transfer function (AASHTO, 2008). Reasonable values of the standard error for each distress transfer function are included in Subsection 2.4 (Distress or Performance Indicator Terms) of this document. The standard errors of the estimate for the IRI regression equations for different pavement types are also included in Section 5 of the *MEPDG Manual of Practice.* A reasonable standard error of the estimate for the IRI is 17 in./mi.

Evaluate the null hypothesis for the sampling template relative to the standard error (refer to Step 6). The null hypothesis for this initial assessment is that there is no significant difference between the standard error for the global and local calibration efforts at the selected level of confidence.

- If the null hypothesis is rejected, there is a significant difference between the standard error terms resulting from use of the global and local calibration values.
  - If the local calibration has a higher standard error term, it is recommended that the distress simulation model be recalibrated in an attempt to lower the standard error—proceed to Step 10. The agency can decide, however, to just accept the higher standard error or default standard error determined from the original calibration process using LTPP test sections. If this is the case, proceed to Step 11.
  - If the local calibration has a lower standard error term, these calibration coefficients are recommended for use—proceed to Step 11.
- If the null hypothesis is accepted, the local and global standard errors are considered the same; these calibration coefficients can be used for pavement design—proceed to Step 11.

Step 10—Reduce Standard Error of the Estimate
If the user decides that the standard error is too large, resulting in overly conservative designs at higher reliability levels, revisions to the local calibration values of the transfer function or statistical model may be needed. This step can be complicated and will probably require external revisions to the local calibration parameters or agency specific values to improve on the prediction model's precision. The following provides some general guidance to accomplish this step.

Step 10.1
Prior to the recalibration or modification process to local conditions, the standard error components should be quantified to estimate the potential reduction in the total standard error term (refer to Section 5). The lack-of-fit or model error is the only portion of the total standard error that can be reduced through the local calibration process, after the hierarchical input level has been selected (refer to Step 1). The measurement error should be quantified and compared to the total error to estimate the potential increase in precision of the prediction model. The measurement error is probably the larger of the error components and making changes to the local calibration (or agency specific) values will not change the magnitude of that error component. The agency needs to decide whether additional costs and effort will significantly reduce the total standard error of the specific distress and IRI prediction models.

- If it is expected that the total standard error cannot be significantly reduced, proceed to Step 11.
- On the other hand, if it is expected that the model precision can be significantly improved, continue with this step.

Step 10.2
Compute the standard error within each block of the sampling template (refer to Section 5) to determine whether the local standard error term is dependent on any primary or secondary tier parameter of the matrix. Results from the analysis of local standard errors within each block can be used to make revisions to specific local calibration parameters.

Step 10.3
Adjust the local calibration values (agency specific values) of the distress transfer functions to reduce the standard error of the recalibration data set. Tables 6-1 and 6-2 list the coefficients of the MEPDG

transfer functions or distress and IRI prediction models that should be considered for revising the predictions to minimize the standard error for flexible and rigid pavements, respectively.

A fitting process of the model constants is evaluated based on a goodness-of-fit criteria on the best set of values for the coefficients of the model. The methods of evaluation make use of either the analytical process for models that suggest linear relationship or make use of numerical optimization for models that suggest non-linear relationship. The analytical approach is based on least squares using multiple regression analysis, stepwise regression analysis, principal components analysis, and/or principal component regression analysis. The numerical optimization includes methods such as the steepest descent or pattern search.

These local calibration values that result in the lowest standard error should be used for pavement design—proceed to Step 11.

Step 11—Interpretation of Results, Deciding on Adequacy of Calibration Parameters

Step 11.1

The local standard error of the estimate for each distress and IRI prediction model should be evaluated to determine the impact on the resulting designs at different reliability levels. The sampling template can be used to determine the design life of typical site features and pavement structures or rehabilitation strategies for different reliability levels. An agency should review the expected pavement/rehabilitation design life within each cell of the sampling template. The agency now has three options to consider.

1. The expected design life is believed to be "reasonable" for the reliability levels used by the agency—not resulting in overly conservative designs based on historical data. For this option or condition, proceed to the last activity; Step 11.2. To define reasonable expected design life, survivability or probability of failure curves should be prepared from the PMS data and compared to the calculated reliability of each segment included in the sampling template.

2. The expected design life is believed to be "too short" for the reliability levels used by the agency, resulting in very conservative and costly designs. The agency should try to reduce the standard error of the estimate for the specific distress simulation model—proceed back to Step 10 for reducing the standard error of the estimate.

3. The expected design life is believed to be "too short" for the reliability levels used by an agency, because the measurement and pure error components are too large relative to the lack-of-fit and input error components. Therefore, making model revisions, adding more validation-calibration roadway segments, using Level 1 input parameters, completing field forensic investigations, and other costly field activities are expected to have a minor impact on the total standard error of the estimate. For this condition, the agency should consider increasing the failure criteria or trigger values for new pavement and rehabilitation designs and proceed to Step 11.2.

Step 11.2

The local calibration values and new standard error of the estimate should be entered into the MEPDG software for use in new pavement and rehabilitation designs.

7.0 Referenced Documents and Standards

7.1 Referenced Documents

AASHTO. *Mechanistic-Empirical Pavement Design Guide—A Manual of Practice.* Publication code: MEPDG-1, ISBN: 978-1-56051-423-7. American Association of State Highway and Transportation Officials, Washington, DC, Interim Edition, July 2008.

FHWA. *Distress Identification Manual for Long-Term Pavement Performance Program* (Fourth Revised Edition). Publication No. FHWA-RD-03-031, Federal Highway Administration, Washington, DC, 2003.

NCHRP. *Refining the Calibration and Validation of Hot Mix Asphalt Performance Models: An Experimental Plan and Database.* NCHRP Results Digest Number 284, National Cooperative Highway Research Program, Transportation Research Board of the National Academies, Washington, DC, December 2003a.

NCHRP, *Jack-Knife Testing—An Experimental Approach to Refine Model Calibration and Validation,* NCHRP Results Digest Number 283, National Cooperative Highway Research Program, Transportation Research Board of the National Academies, Washington, DC, December 2003b.

NCHRP. Changes to the Mechanistic-Empirical Pavement Design Guide Software Through Version 0.900. *NCHRP Research Results Digest 308,* National Cooperative Highway Research Program, Transportation Research Board of the National Academies, Washington, DC, September 2006.

NCHRP. *Version 1.0—Mechanistic-Empirical Pavement Design Guide Software.* National Cooperative Highway Research Program, National Academy of Sciences, Washington, DC, April 2007.

Rada, G. R., et al. *Study of LTPP Distress Variability,* Volume 1. Publication No. FHWA-RD-99-074, Federal Highway Administration, Office of Infrastructure Research and Development, McLean, Virginia, September, 1999.

7.2 Test Protocols and Standards

AASHTO R 55 *Standard Practice for Quantifying Cracks in Asphalt Pavement Surfaces*

AASHTO R 57 *Standard Practice for Operating Inertial Profilers and Evaluating Pavement Profiles*

AASHTO R 36 *Standard Practice for Evaluating Faulting of Concrete Pavements*

AASHTO R 48 *Standard Practice for Determining Rut Depth in Pavements*

AASHTO T 256 *Standard Method for Pavement Deflection Measurements*

Appendix: Examples and Demonstrations for Local Calibration

A1. Background

All performance indicator prediction models in the *Mechanistic-Empirical Pavement Design Guide* (MEPDG) were calibrated to observed field performance from a representative sample of pavement test sites located throughout North America. These models are defined as being globally calibrated. Data from the Long-Term Pavement Performance (LTPP) test sections were used extensively in the global calibration process, because of their consistency in the monitored data over time and the diversity of test sections spread throughout North America. Other experimental test sections, such as MnRoad and Vandalia, were also included in the global calibration process.

Policies on pavement preservation and maintenance, construction and material specifications, and design features vary across the United States and are not considered directly in the MEDPG. These factors can be considered indirectly through the local or agency specific calibration coefficients included in the MEPDG, if found to cause bias in the performance predictions. The purpose of this appendix is to provide examples using the *Local Calibration Guide* for validating and/or revising the MEPDG global calibration factors to account for local conditions and materials that were not considered in the global calibration process.

It is impossible to cover all conditions and scenarios that user agencies may encounter in validating the acceptability of the global prediction models. The objective of this appendix is to provide a listing of the steps, with examples of judgments and decision-making criteria, in deciding whether to accept the global calibration factors or develop local calibration or agency specific factors in predicting the performance indicators.

This appendix is divided into three parts, including this Background. Appendix Two (A2) is focused on flexible pavements and common rehabilitation strategies for flexible pavements (hot mix asphalt [HMA] overlays), while Appendix Three (A3) is focused on rigid pavements (specifically, new jointed plain concrete pavements). All demonstrations were focused on validating and/or revising the global calibration coefficients for the common design strategy used by two agencies (the Kansas and Missouri Departments of Transportation).

Each pavement type uses two data sources for demonstrating the local validation-calibration effort—segments within a pavement management system (PMS) and LTPP test sections. Both data sets are used to demonstrate the steps and potential differences between measures of distress that are consistent and inconsistent with the definitions used during the global calibration effort under the NCHRP studies. These two different sets for the flexible pavements and HMA overlays were also used to demonstrate differences in the quality and quantity of the input data.

The PMS segments used in the examples include those from the Kansas Department of Transportation (KSDOT) for flexible pavements and the Missouri Department of Transportation (MODOT) for rigid pavements. The KSDOT segments are focused on their common design strategy or full-depth flexible pavements and HMA overlays, while the MODOT segments are focused on plain jointed concrete pavements (JPCP). The LTPP SPS-1 and SPS-5 experiments are used for flexible pavements, while the SPS-2 experiment is used for rigid pavements.

Another objective of this appendix is to demonstrate how data from different sources (PMS and LTPP databases) can be combined to establish regional, as well as local calibration factors for each performance measure. The *Guide for Local Calibration* is used in determining the local calibration values from each data set and pavement type.

A2. New Flexible Pavements and Rehabilitation of Flexible Pavements

A2.1 Demonstration 1—PMS Data and Local Calibration

A2.1.1 Description of PMS Segments

Sixteen projects from the KSDOT PMS database were selected for demonstrating use of the *Local Calibration Guide*. Eleven of the projects were HMA full-depth new construction/reconstruction and six were HMA overlays of flexible pavements. Table A2-1 provides general information about the pavement structures, while Table A2-2 provides project descriptions and their locations. Figure A2-1 shows the general location of each PMS segment. Table A2-3 summarizes the material types and layer thicknesses that were extracted from construction files. The binder types used in the projects included conventional or neat HMA, polymer modified asphalt (PMA), and Superpave mixtures. More detailed descriptions of each project are provided in Attachment A2.4.A.

A2.1.2 Step-by-Step Procedure

The steps included in the *Local Calibration Guide* were followed for this demonstration using PMS data. Many of the decisions made by KSDOT were based on an expedited time frame to collect the necessary data for the demonstration. KSDOT would likely make different decisions given a longer time frame for the local calibration process.

Table A2-1. General Structure Information for the Selected Kansas PMS Projects

No.	Payment Type	Project ID (KSDDOT PMS ID)	Binder Type	Number and Length of Homogenous Sections	Construction and Maintenance and Rehabilitation History	Comments	
						Initial IRI (in./mi)	Analysis Period (yr)
1	Full-Depth HMA	FDAC-C-2 (0673047017180)	Conventional HMA Mix	1 section 0.94 mi	1992. Reconstruction. (1" HMA + 8" Recycle HMA + 6" Lime subgrade + subgrade) 2001. Overlay (1.6" HMA overlay)	85	1993–2001
2		FDAC-C-4 (0223120008080)	Conventional HMA Mix	1 section 0.83 mi	1989. Reconstruction. (1.5" HMA + 6.5" HMA + subgrade) 1999. Overlay (1.6 HMA overlay)	110	1990–1999
3		FDAC-P-1 (0552083020210)	PMA Mix	1 section 0.91mi	1999. Reconstruction. (1" HMA + 4" HMA + 8.7" HMA + 4" Lime subgrade + subgrade)	36	2000–2006
4		FDAC-P-2 (0552083026270)	PMA Mix	1 section 0.50 mile	1999. Reconstruction. (1" HMA+4" HMA + 8.7"HMA + 4" Lime subgrade + subgrade)	35	2000–2006
5		FDAC-P-3 (0552083028290)	PMA Mix	2 sections: I: 0.34 mi II: 0.13 mi	Section I: 1999. Reconstruction. (1.0" HMA + 4.0" HMA + 4.7" HMA + 3.5" Lime subgrade + subgrade Section II: 1999. Reconstruction. (1.0" HMA + 4.0" HMA+ 8.7" HMA + 4" Lime subgrade + subgrade	40.5	2000–2006
6		FDAC-P-4 (0552083021220)	PMA Mix	1 section 0.45 mi	1999. Reconstruction. (1" HMA + 4" HMA + 8.7" HMA + 4" Lime subgrade + subgrade)	39.0	2000–2006
7		FDAC-P-5 (0552083023240)	PMA Mix	1 section 0.38mi	1999. reconstruction (1" HMA + 4" HMA + 8.7"HMA + 4" Lime subgrade + subgrade)	37.0	2000–2006
8		FDAC-S-1 (0372054012120)	Superpave	1 section, 0.67 mi	1998. Reconstruction. (1" HMA + 15" HMA + 6" Lime subgrade + subgrade)	44.0	1999–2006
9		FDAC-S-3 (0702075007080)	Superpave Mix	2 sections, same structure—0.84 mi	1998 Reconstruction. (1" HMA + 4" HMA + 6" HMA + 6" Lime subgrade + subgrade)	55.0	1999–2006

Continued on next page

Table A2-1—Continued

No.	Payment Type	Project ID (KSDDOT PMS ID)	Binder Type	Number and Length of Homogenous Sections	Construction and Maintenance and Rehabilitation History	Comments: Initial IRI (in./mi)	Comments: Analysis Period (yr)
10	Full-Depth HMA	FDAC-S-4 (0632075032330)	Superpave Mix	1 section 0.81mi	1998. Reconstruction. (1" HMA + 12" HMA + 6" Lime subgrade + subgrade)	54.0	1999–2006
11		FDAC-S-5 (0702075008090)	Superpave Mix	1 section 4.2 mi	1998 Reconstruction (1" HMA + 4" HMA + 6" HMA + 6" Lime subgrade + subgrade)	40.0	1999–2006
12	HMA Overlay	HMA_HMA-C-1 (0083177008090)	Conventional HMA Mix	1 section 1.39 mi	• 1979 New construction. (1" HMA + 7" HMA + 8" Lime subgrade + subgrade) • 1989. Overlay (¾" overlay + 1" recycled HMA) • 1997. Overlay (3" HMA Overlay)	70.5	1998–2005
13		HMA_HMA-C-2 (0083177006070)	Conventional HMA Mix	1 section 2.16 mi	• 1979 New construction. (1" HMA + 5" HMA + 8" Lime subgrade + subgrade) • 1989. Overlay (¾" overlay + 1" recycle HMA) • 1997. Overlay (3" HMA Overlay)	61.5	1998–2006
14		HMA_HMA-C-4 (0132160005060)	Conventional HMA Mix	1 section 0.29 mi	• 1964 New construction. (7.5" HMA + subgrade) • 1996. Overlay. (1" + 4.5" HMA overlay)	37.0	1997–2006
15		HMA_HMA-P-3 (0233010007080)	PMA Mix	1 section 1.10 mi	• 1996. Reconstruction. (1" HMA + 8"HMA + 6" Lime subgrade + subgrade) • 2000 Overlay. (1.6" + 2.0" HMA overlay)	49.5	2001–2006
16		HMA_HMA-S-3 (0052281018200)	Superpave Mix	2 sections I: 0.04 mi II: 0.78 mi	Section II: • 1965 New Construction. (1.5" + 3.0" HMA + subgrade) • 1999. Overlay (1.5" HMA + 4.0" recycled HMA) • 2005. Overlay (1" overlay superpave)	53.5	2000–2005

Table A2-2. General Project Information for the Kansas PMS Segments

Project Name	KSDOT PMS ID	Length (mi)	Direction	Route	Begin Milepost	End Milepost	County Number	County Name	District Number
FDAC-C-2	0673047017180	0.94	Northbound	State 47	32.913	33.855	67	Neosho	4
FDAC-C-4	0223120008080	0.83	Eastbound	State 120	8.098	8.933	22	Doniphan	1
FDAC-P-1	0552083020210	0.91	Northbound	US 83	144.713	145.625	55	Logan	3
FDAC-P-2	0552083026270	0.50	Northbound	US 83	150.360	150.857	55	Logan	3
FDAC-P-3 (Section I)	0552083028290	0.34	Northbound	US 83	153.092	153.436	55	Logan	3
FDAC-P-3 (Section II)	0552083028290	0.13	Northbound	US 83	153.436	153.566	55	Logan	3
FDAC-P-4	0552083021220	0.45	Northbound	US 83	145.625	146.075	55	Logan	3
FDAC-P-5	0552083023240	0.38	Northbound	US 83	148.185	148.566	55	Logan	3
FDAC-S-1	0372054012120	0.67	Eastbound	US 54	282.514	283.193	37	Greenwood	4
FDAC-S-3	0702075007080	0.84	Northbound	US 75	120.227	121.068	70	Osage	1
FDAC-S-4	0632075032330	0.81	Northbound	US 75	32.679	33.493	63	Montgomery	4
FDAC-S-5	0702075008090	4.21	Northbound	US 75	121.068	125.275	70	Osage	1
HMA_HMA-C-1	0083177008090	1.39	Northbound	State 177	7.414	8.807	8	Butler	5
HMA_HMA-C-2	0083177006070	2.16	Northbound	State 177	5.255	7.414	8	Butler	5
HMA_HMA-C-4	0132160005060	0.29	Eastbound	US 160	135.261	135.550	13	Clark	6
HMA_HMA-P-3	0233010007080	1.10	Eastbound	State 10	7.332	8.430	23	Douglas	1
HMA_HMA-S-3 (Section I)	0052281018200	0.04	Northbound	US 281	115.497	115.539	5	Barton	5
HMA_HMA-S-3 (Section II)	0052281018200	0.78	Northbound	US 281	115.539	116.315	5	Barton	5

Table A2-3. Material Types and Layer Thickness for trhe Kansas PMS Segments

Project ID	Layer Information			Initial AADTT; Two Way	Climate	
	No.	Material Type	Thickness, in.		Latitude	Longitude
FDAC-P-2	1	Fine-Grained Soil	—	218	39.07	−100.85
	2	Lime Modified Soil	4			
	3	Neat HMA Base Mixture	8.7			
	4	PMA Binder and Wearing Surface	5			
FDAC-P-3	1	Fine-Grained Soil; A-7-6	—	210	39.10	−100.85
	2	Lime Modified Soil	3.5			
	3	Neat HMA Base Mixture	4.7			
	4	PMA Binder and Wearing Surface	5			
FDAC-P-4	1	Fine-Grained Soil; A-7-6	—	148	39.00	−100.85
	2	Lime Modified Soil	4			
	3	Neat HMA Base Mixture	8.7			
	4	PMA Binder and Wearing Surface	5			
FDAC-P-5	1	Fine-Grained Soil; A-7-6	—	156	39.00	−100.85
	2	Lime Modified Soil	4			
	3	Neat HMA Base Mixture	8.7			
		PMA Binder and Wearing Surface	5			
FDAC-S-1	1	Fine-Grained Soil; A-7-6	—	155	37.81	−96.31
	2	Lime Modified Soil	6			
	3	Superpave HMA Mixtures	16			
FDAC-S-3	1	Fine-Grained Soil; A-7-6	—	114	38.52	−95.70
	2	Lime Modified Soil	6			
	3	Superpave HMA Base Mixture	6			
	4	Superpave Binder and Wearing Course	5			
FDAC-S-4	1	Fine-Grained Soil; A-7-6	—	129	37.34	−95.71
	2	Lime Modified Soil	6			
	3	Superpave HMA Mixtures	13			
FDAC-S-5	1	Fine-Grained Soil; A-7-6	—	99	38.54	−95.69
	2	Lime Modified Soil	6			
	3	Superpave Base Mixture	6			
	4	Superpave Binder and Wearing Course	5			
HMA_ HMA-C-1	1	Fine-Grained Soil; A-7-6	—	7	37.92	−96.75
	2	Lime Modified Soil	8			
	3	HMA Mixture—Existing	6.75			
	4	Neat HMA Overlay	4			

Table A2-3—*Continued*

Project ID	Layer Information			Initial AADTT; Two Way	Climate	
	No.	Material Type	Thickness, in.		Latitude	Longitude
HMA_ HMA-C-2	1	Fine-Grained Soil; A-7-6	—	7	37.92	−96.75
	2	Lime Modified Soil	8			
	3	HMA Mixture—Existing	6.75			
	4	Neat HMA Overlay	3			
HMA_ HMA-C-4	1	Fine-Grained Soil; A-7-6	—	64	37.29	−100.01
	2	HMA Mixture—Existing	7.5			
	3	Neat HMA Overlay	5.5			
HMA_ HMA-P-3	1	Fine-Grained Soil; A-7-6	—	100	38.92	−95.29
	2	Lime Modified Soil	6			
	3	HMA Mixture—Existing	9			
	4	Neat HMA Overlay	3.6			
HMA_ HMA-S-3	1	Fine-Grained Soil; A-7-6	—	61	38.52	−98.71
	2	HMA Mixture—Existing (recycled)	5.5			
	3	Neat HMA Overlay	1.5			

Step 1—Select Hierarchical Input Level

The hierarchical input level to be used in the local validation-calibration process should be consistent with the way the agency intends to determine the inputs for day-to-day use. This demonstration using PMS roadway segments is for the condition for which minimum data are available.

Input Levels 2 and 3 were used for all input parameters for the PMS segments—most were input Level 3. Data needed to determine Level 1 inputs were unavailable for the PMS segments. The general information from which the inputs were determined for each input category is discussed in Step 5.

Step 2—Experimental Factorial and Matrix or Sampling Template

Creating a detailed sampling template was not completed within this example, because only two pavement cross sections or design strategies were used within this example; full-depth HMA and HMA overlays of flexible pavements. It was decided that the global calibration values would be validated for the HMA design strategies and materials commonly used in Kansas. Table A2-4 shows the simplified sampling template used for this demonstration, along with the number of PMS segments or projects within each cell (refer to Step 4).

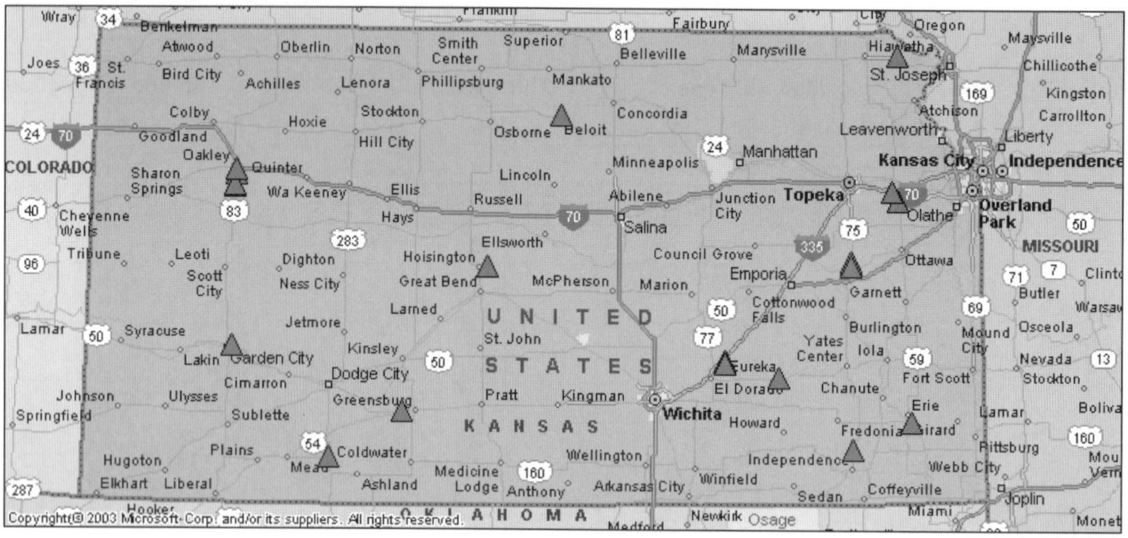

Figure A2-1. General Location of the Roadway Segments Selected for Demonstrating the Local Validation-Calibration Process Using Kansas PMS Data

Table A2-4. Simplified Sampling Template for the Demonstration Using PMS Data

HMA-Mixture Type	Pavement Type		Total PMS Segments
	Full-Depth Reconstruction	HMA Overlays	
Conventional Neat HMA Mixtures	2	3	5
Superpave Mixtures	4	1	5
PMA Mixtures	5	1	6
Total PMS Segments	11	5	16

Extreme climate variations and soil conditions do not occur across Kansas, with the exception of some localized areas. The only other primary tier in the factorial is HMA mixture type—neat HMA, PMA, and Superpave mixtures. KSDOT has adopted Superpave and has been moving towards the use of PMA mixtures to reduce rutting and thermal cracking since the late 1990s. Thus, pavement type (full-depth and HMA overlays) and HMA mixture type were the only two tiers of the sampling template for the PMS segments.

The number of roadway segments selected for the sampling template should result in a balanced factorial with the same number of replicates within each cell. As shown in Table A2-4, the factorial is unbalanced and some cells do not have replication.

Step 3—Estimate Sample Size for Each Performance Indicator Prediction Model

A 90-percent level of significance was used for estimating the sample size (total number of roadway segments or projects). Higher confidence levels can be used, but that will increase the number of segments needed. Table A2-5 summarizes the estimated sample size for each measure of performance. A minimum of four observations per project was assumed. The number of distress observations per segment is dependent on the measurement error or within segment data variability over time (i.e., the higher the within project data dispersion or variability, the larger the number of observations needed for each distress). The number of distress measurements made within a roadway segment is also dependent on the within project variability of the design features and site conditions.

Table A2-5. Estimated Number of PMS Segments Needed for the Local Validation-Calibration Process

Performance Indicator	Design Criteria or Magnitude	Tolerable Bias, e_t	s_e/s_y	Minimum Number of Samples	
				Projects	Observations
Rutting, in.	0.40	0.05	0.50	8	32
Fatigue Cracking, %	20	2.5	0.50	8	32
Thermal Cracking, ft/mi	1,500	300	0.50	18	72
Roughness, in./mi	130	10	0.50	12	48

$$n = \left(\frac{Z_{\alpha/2}\sigma}{E} \right)^2$$

where:
z_α = 1.282 for a 90 percent confidence interval;
s_y = standard deviation of the maximum true or observed values; and
e_t = tolerable bias.

The tolerable bias was estimated from the levels that are expected to trigger some major rehabilitation activity (see Table A2-5), which are agency dependent. The s_e/s_y value (ratio of the standard error and standard deviation of the measured values) will also be agency dependent. A value of 0.50 was selected for this demonstration, which is fairly low for PMS data.

Step 4—Select Roadway Segments
Projects should be selected to cover a range of distress values that are of similar ages within the sampling template. Roadway segments exhibiting premature or accelerated distress levels, as well as those exhibiting superior performance (low levels of distress over long periods of time), can be used, but with caution. The roadway segments selected for the sampling template when using hierarchical input Level 3 should represent average performance conditions.

A limited number of potential PMS segments were available for this example. Many of the PMS segments had insufficient construction histories or insufficient distress data to be included in the local validation-calibration procedure.

Sixteen segments with multiple distress measurements within each segment were selected for this demonstration (see Figure A2-1 and Table A2-2). The 16 segments, however, are believed to be sufficient for the two pavement structures and different surface mixtures considered within the demonstration (see Tables A2-4 and A2-5). It is important that the same number of performance observations per age per project be available in selecting roadway segments for the sampling template. It would not be good practice to have some segments with ten observations over 10 years with other segments having only two or three observations over 10 years. The segments with one observation per year would have a greater influence on the validation-calibration process than the segments with 0.25 observations per year. The number of observations per year for the 16 PMS segments selected vary from 1.0 observation per year for the full-depth reconstruction projects to about 0.75 observations per year for the HMA-Overlay Projects. This range is considered acceptable.

Step 5—Extract and Evaluate Distress and Project Data

This step is grouped into four activities: (1) extracting and reviewing the performance data; (2) comparing the performance indicator magnitudes to the trigger values; (3) evaluating the distress data to identify anomalies and outliers; and (4) determining the inputs to the MEPDG.

Step 5.1—Extract, Review, and Convert Measured Values to the Values Predicted by the MEPDG, if Needed.

First, the distress or performance indicator measurements included in the KSDOT PMS database were reviewed to determine whether the measured values are consistent with the values predicted by the MEPDG. For the KSDOT PMS data, the measured cracking values are different, while the rutting and IRI values are similar and assumed to be the same. The cracking values and how they were used in the local calibration process are defined below.

- **Fatigue Cracking**. KSDOT measures fatigue cracking in number of wheel path feet per 100-ft sample by crack severity, but does not distinguish between alligator cracking and longitudinal cracking in the wheel path. In addition, reflection cracks are not distinguished separately from the other cracking distresses. The PMS data were converted to a percentage value similar to what is reported in the HPMS system from Kansas. In summary, the following equation was used to convert KSDOT cracking measurements to a percentage value that is predicted by the MEPDG.

$$FC = \left(\frac{FCR_1(0.5) + FCR_2(1.0) + FCR_3(1.5) + FCR_4(2.0)}{8.0} \right) \quad (A2\text{-}1)$$

All load related cracks are included in one value. Thus, the MEPDG predictions for load related cracking were combined into one value by simply adding the length of longitudinal cracks and reflection cracks for HMA overlays, multiplying by 1.0 ft, dividing that product by the area of the lane and adding that value to the percentage of alligator cracking predicted by the MEPDG.

- **Transverse Cracking**. Another difference is that KSDOT records thermal or transverse cracks as the number of cracks by severity level. The following equation has been used by KSDOT to convert their measured values to the MEPDG predicted value of ft/mi.

$$TC = \left(\frac{TCR_o + TCR_1 + TCR_2 + TCR_3}{(10)(12)(52.8)} \right) \quad (A2\text{-}2)$$

The value of 10 in the above equation is needed because the data are stored with an implied decimal. The value of 12 is the typical lane width, and the value of 52.8 coverts from 100-ft sample to a per mi basis. Prior to 1999, KSDOT did not record the number or amount of sealed transverse cracking (TCR_0). As a result, the amount of transverse cracks sometimes goes to "0".

The average measured value should be determined for each measurement period for each PMS segment. The time-history data should not be smoothed but can be cleaned to remove data errors. As an example, measured distresses that are recorded as zero after multiple values of distress have been recorded. Plots of the average time-history data are included in Attachment A2.4.B. Some important observations of the data that have an impact on the validation-calibration process are listed below.

- Large measurement errors are present for all performance indicators. The measured values significantly increase and decrease with time. In addition, there are abrupt changes in the rutting and IRI data over time. Thus, improving on the precision of the prediction model is not likely.

- Few of the PMS segments have any measured fatigue cracking. Thus, confirming the fatigue cracking global calibration values is not likely.

- Rut depths are low for all PMS segments included within this demonstration. Thus, confirming the rut depth global calibration values will be limited to rut depths significantly less than the design criteria or trigger value.

Step 5.2—Compare Distress Magnitudes to Trigger Values

The next activity of this step is to compare the distress magnitudes to the trigger values for each distress. Then answer the question—Does the sampling template include values close to the design criteria or trigger value? Table A2-6 summarizes the average, maximum, and minimum distress values for each performance indicator as compared to the trigger values (design criteria) for some major rehabilitation activity (see Table A2-5).

Table A2-6. Summary of the Maximum Values of Different Performance Indicators in Comparison to the Design Criteria or Trigger Values (Number of Sites = 16)

Distress or Performance Indicator	Design Criteria	Maximum Values Measured for Each Segment				Probability of Exceeding Trigger Value, %
		Average Max. Value	Lowest Max. Value	Largest Max. Value	Stand. Dev. of Max. Values	
Rut Depth, in.	0.40	0.22	0.17	0.36	0.0754	0.8
Fatigue Cracking, %	20	2.7	0	20.8	4.44	0.0
Transverse Cracks, ft/mi	1,500[1]	969	0	4,689	1,237	33.4
Roughness, in./mi	130	87	55	154	26.9	5.5

Note 1: The 1,500 ft/mi corresponds to an average crack spacing of about 40 ft.

As tabulated, most of the observed or measured distress values are significantly less than the design criteria. In fact, the average distress magnitudes are more than two standard deviations below the design criteria, with the exception of transverse cracking. Table A2-6 also summarizes the expected probability that the trigger values will be exceeded using the PMS local calibration data set. As shown, the probability of exceeding the trigger values is low for all performance indicators, with the exception of transverse cracking.

This comparison suggests that the values used as KSDOT's trigger values are too high or the flexible pavements and HMA overlays are being rehabilitated for other reasons. This observation becomes important when evaluating the bias and standard error terms of the prediction models under Steps 7 and 9, respectively. More importantly, the maximum area of fatigue cracking measured is less than 3 percent. This level of fatigue cracking is too low to validate and accurately determine the local calibration values or adjustments for predicting the increase in cracking over time; especially when 20 percent fatigue cracking was selected for the design criteria.

Step 5.3—Evaluate Distress Data to Identify Anomalies and/or Outliers

The distress data for each roadway segment included in the sampling template should be reviewed prior to determining all of the MEPDG input parameters. This evaluation can be limited to visual inspection of the data over time to ensure that the distress data are reasonable—time-history plots or include a detailed statistical comparison of the performance data. As a minimum, the following questions should be asked (Attachment A2.4.B includes graphs that show the distress values over time for the roadway segments).

- Does the data make sense within and between each roadway segment? All of the data extracted from the Kansas PMS segments looked reasonable and appears to represent typical performance characteristics and conditions. Obviously, any zeros that represent non-entry values should be removed from the local validation-calibration database. Distress data that return to zero values within the measurement period may indicate some type of maintenance or rehabilitation activity.
- Measurements taken after structural rehabilitation should be removed from the database or the observation period should end prior to the rehabilitation activity.
- Distress values that are zero as a result of some maintenance or pavement preservation activity, which is a part of the agency's management policy, should be removed but future distress observation values after that activity should be used.

- Are there segments with anomalies, outliers, or blunders in the data? If the outliers or anomalies can be explained and are a result of some non-typical condition, they should be removed. If the outlier or anomaly cannot be explained, they should remain in the database. No outliers or anomalies were found in the KSDOT data. The magnitude of the performance indicators, however, do increase and then decrease exhibiting high within project variability in the measured values (refer to Attachment A2.4.B). The number of measurements per segment was increased from four to a minimum of 10, because of the high within segment variability.

Step 5.4—Inputs to the MEPDG for Each Input Category

The following provides a brief discussion on the information extracted from the KSDOT databases and files for determining the inputs needed to execute the MEPDG for each PMS segment.

- **Initial IRI**—As noted in the above project descriptions, the initial IRI was determined from the measured values within one or two years after construction. This value is believed to be reasonable, because only minor magnitudes of distress were recorded for the first couple of years after construction.

- **Construction Histories, Cross Sections, and Layer Thicknesses**—As-built plans that were available from KSDOT records were used to determine the material types and layer thicknesses for each PMS segment. It was assumed that all layers were placed in accordance with the project specifications. Material properties consistent with the specifications were used for inputs with minimum construction data.

  - The construction date for the full-depth pavements and HMA overlays was obtained from the as-built plans or construction database files.
    - For all full-depth reconstruction projects, it was assumed that the first lift of HMA was placed one month following preparation of the subgrade and that date was entered as the construction date.
    - For the HMA-Overlay Projects, the construction date or time that the first overlay lift was placed was assumed to be the start of construction. The construction date of the existing flexible pavement for the HMA-Overlay Projects was taken from the construction database.
    - The opening month to traffic for all projects was assumed to be one month following the HMA placement.

- **Rehabilitation Inputs**—The condition of the HMA pavement prior to overlay was determined from the distress values included in the PMS database prior to overlay. All layers were assumed to be fully bonded.

- **Traffic**—Default values were used for all input with the exception of speed, number of lanes, traffic growth, truck traffic classification groups, and average annual daily truck traffic (AADTT).
  - The posted speed limit was used as the input for all PMS segments.
  - The AADTT was taken from the KSDOT traffic database for each roadway segment.
  - The AADTT or annual equivalent single axle loads (ESALs) included in KSDOT's database were used to estimate the average growth in truck traffic for each PMS segment.

- **Climate**—The longitude and latitude of each PMS segment was used to create a virtual weather station for that segment of roadway. The weather stations in Kansas and adjacent states were used to create the virtual weather stations.

- **Materials**—The material and layer properties for each pavement layer and subgrade were taken from construction records, when available. If adequate data were unavailable, the mean value from the specifications was used or the average value determined for the specific input from other projects with similar material.
  - Dynamic modulus data were unavailable for all HMA mixtures and resilient modulus data were unavailable for all soils. Thus, Level 3 or the default values included in the MEPDG were used in all cases.
  - The creep compliance and indirect tensile strength for all HMA mixtures were unavailable for all projects. Thus, the default values (Input Level 3) were used in all cases.

Step 6—Conduct Field and Forensic Investigations
If the assumptions in the MEPDG are challenged by the agency, forensic investigations are needed to measure the rutting in the individual layers and to determine where the cracks initiated or the direction of crack propagation. Steps 7 through 11 are executed for each specific MEPDG predicted distress after the forensic investigations are completed.

For this demonstration, KSDOT decided to accept the assumptions and conditions included in the MEPDG for the global calibration effort. Thus, no field and forensic investigations were planned or conducted to determine the location of crack initiation and the amount of rutting within each pavement layer and foundation.

Step 7—Assess Local Bias from Global Calibration Factors

The MEPDG was executed using the global calibration values to predict the performance indicators for each PMS segment. The predicted performance measures are shown in Attachment A2.4.B relative to the measured values for the PMS segments. The null hypothesis is first checked for the entire sampling matrix. The null hypothesis is that the average residual error or bias is zero for a specified confidence level or level of significance. A 90-percent confidence level was used in this demonstration.

$$H_O : \sum_{i=1}^{n} (y_{Measured} - x_{Predicted})_i = 0 \qquad (A2\text{-}3)$$

Table A2-7 lists the bias for each performance indicator for the entire sampling template and whether the null hypothesis is rejected or accepted, while Figures A2-2 to A2-5 compare the predicted and measured values for each performance indicator. Figures A2-2 and A2-5 also include a comparison between the residual errors (e_r) and predicted values (x_i). This same comparison was excluded from Figures A2-3 and A2-4, because no to nil fatigue and transverse cracks were predicted for most of the PMS segments that exhibited measurable cracks.

Table A2-7. Summary of the Statistical Parameters—Global Calibration Values Used for Predicting Performance Indicators for the Kansas PMS Sections

Performance Indicator	Project	Bias, $e_{r\,(Mean)}$	Standard Error, s_e	s_e/s_y	R^2 (see Note)	Hypothesis; H_o: $y_i - x_i = 0$	Comment
Rutting	New	–0.00412	0.1004	1.06	Poor	Accept; $p = 0.484$	Extensive variability.
	Rehab.	+0.0352	0.0767	1.33	Poor	Accept; $p = 0.269$	Extensive variability; over predicting rut depths.
Fatigue Cracking	New	–0.807	2.65	1.00	Poor	Accept; $p = 0.302$	Limited fatigue cracking was predicted and only a few sections exhibited fatigue cracks. Insufficient number of sections to complete calibration process. No fatigue cracks were predicted for any of the HMA overlays.
	Rehab.	–0.657	1.46	0.948	Poor	Accept; $p = 0.334$	

Table A2-7—*Continued*

Performance Indicator	Project	Bias, $e_{r\,(Mean)}$	Standard Error, s_e	s_e/s_y	R^2 (see Note)	Hypothesis; H_o: $y_i - x_i = 0$	Comment
Transverse Cracking	New	−64.4	702.6	1.40	Poor	Accept; p = 0.413	Thermal cracks were predicted for only two PMS segments, while high amounts of transverse cracks were exhibited on multiple segments. No transverse cracks were predicted for any of the HMA overlays.
	Rehab.	−448.0	685.3	1.02	Poor	Accept; p = 0.252	
IRI	New	−9.65	26.15	0.927	0.22	Accept; p = 0.367	The hypothesis should be checked and evaluated after any bias has been reduced for the other distresses.
	Rehab.	−8.13	27.60	1.66	Poor	Accept; p = 0.312	

Note: Poor means that the model did not explain variation in the measured data within and between the PMS segments.
Residual Error = $e_r = y_i - x_i$
y_i = Measured or Observed Value; Standard Deviation of the observed values.
x_i = Predicted Value

As shown in Table A2-7, the hypothesis is accepted for the transfer functions using all of the PMS segments included within the sampling template. The reason that the hypothesis was accepted is that the bias is low in comparison to the mean measured value and within project variability of the measured distress values. Two other model parameters, however, were used to evaluate model bias—the intercept (b_o) and slope (m) estimators using the following fitted linear regression model between the measured (y_i) and predicted (x_i) values.

$$\hat{y}_i = b_o + m(x_i) \qquad (A2\text{-}4)$$

A non-linear regression model could also be considered and used to reduce the standard error between the predicted mean measured values and predicted values. If the predicted mean values ($\hat{y}_i$) fall along the line of equality, b_o = 0.0, m = 1.0, and $\hat{y}_i = x_i$.

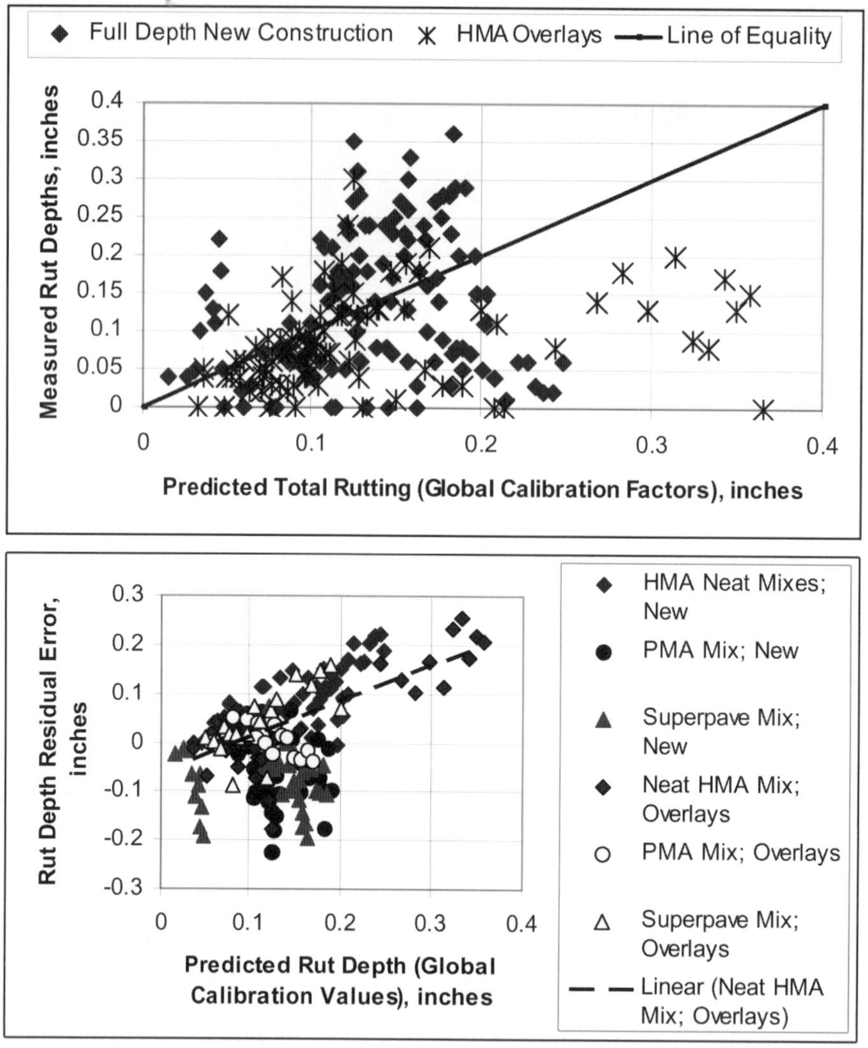

Figure A2-2. Comparison of Predicted and Measured Rut Depths Using the Global Calibration Values and Local Calibration Values of Unity

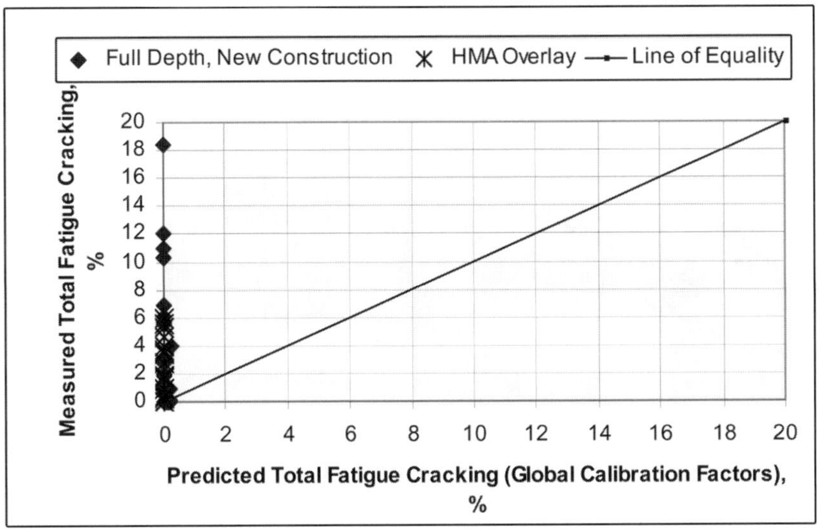

Figure A2-3. Comparison of Predicted and Measured Fatigue Cracking Using the Global Calibration Values and Local Calibration Values of Unity

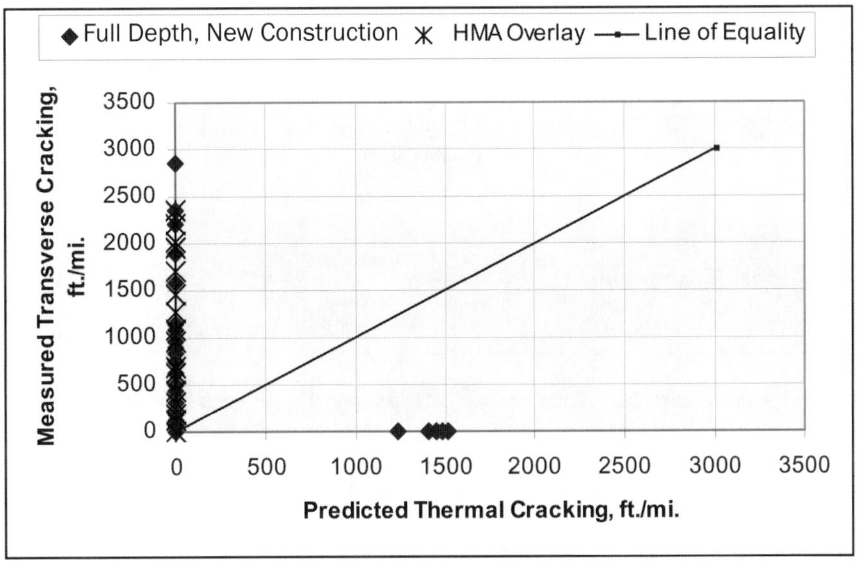

Figure A2-4. Comparison of Predicted Thermal Cracking and Measured Transverse Cracking Using the Global Calibration Values and a Local Calibration Value of Unity

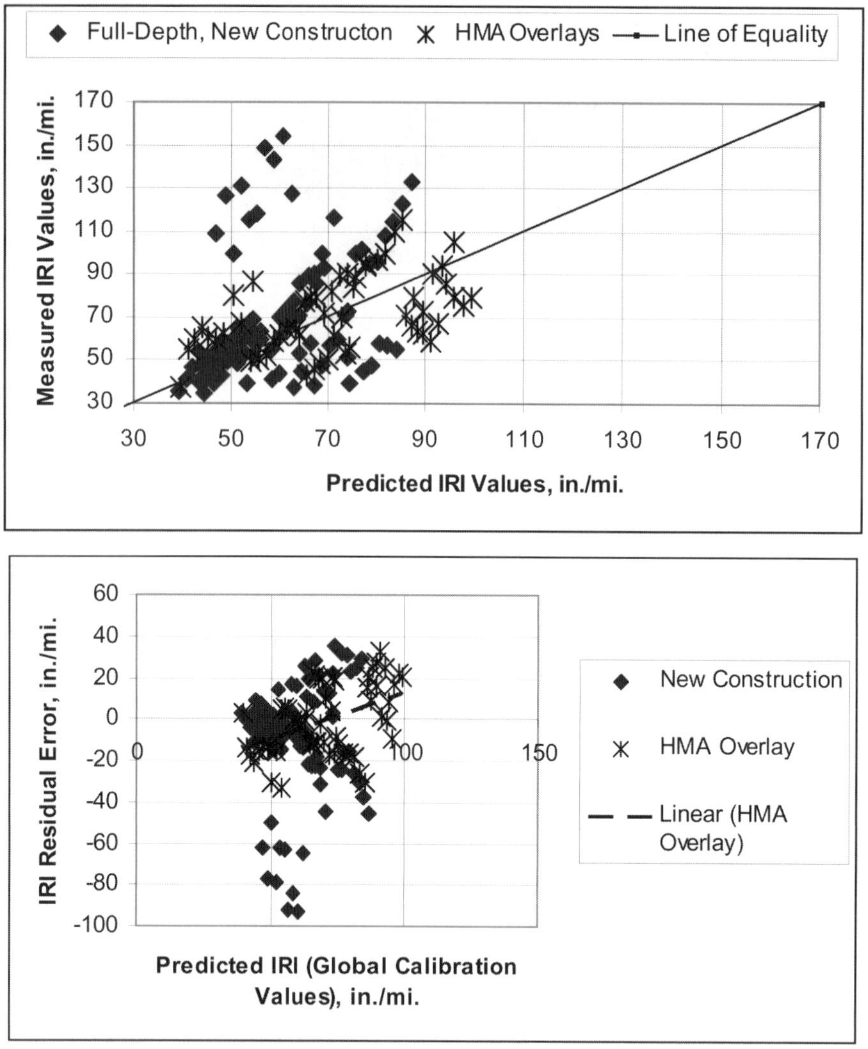

Figure A2-5. Comparison of Predicted and Measured IRI Using the Global Calibration Values and Local Calibration Values of Unity

Figures A2-6 and A2-7 show examples of using these estimators of the mean measured values for rut depth and IRI for the primary cells of the sampling template. As shown, the intercept estimator is significantly different from 0, and/or the slope estimator is significantly different from 1.0. More importantly, the intercept and/or slope estimators are dependent on the primary tiers of the sampling template. In summary, all transfer functions exhibited similar trends or bias and that bias is related to pavement structure (new construction versus rehabilitation) and mixture type (conventional neat HMA versus PMA versus Superpave mixtures).

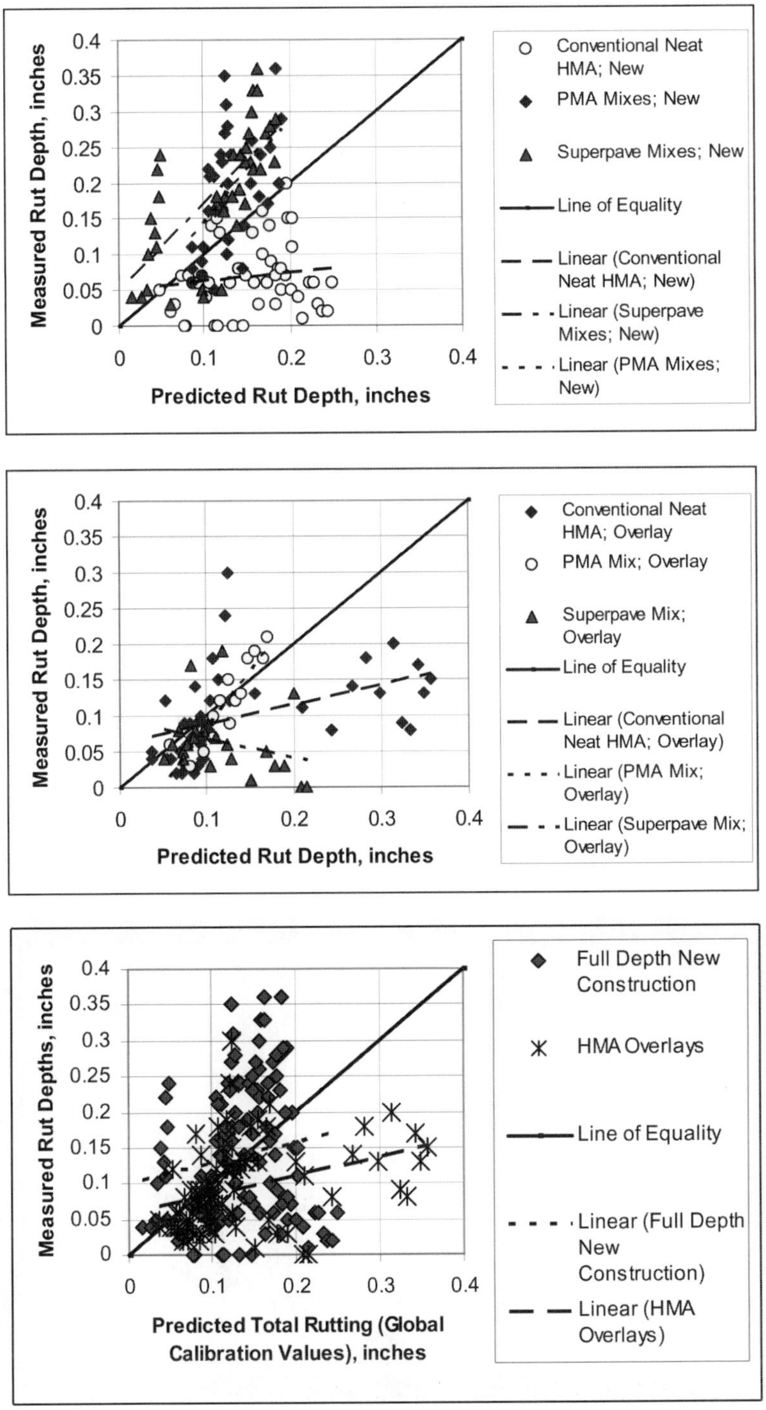

a. Intercept and slope estimators that are dependent on mixture type for the new construction PMS segments.

b. Intercept and slope estimators that are dependent on mixture type for the rehabilitation PMS segments.

c. Intercept and slope estimators that are structure dependent for the PMS segments.

Figure A2-6. Comparison of the Intercept and Slope Estimators to the Line of Equality for the Predicted and Measured Rut Depths Using the Global Calibration Values

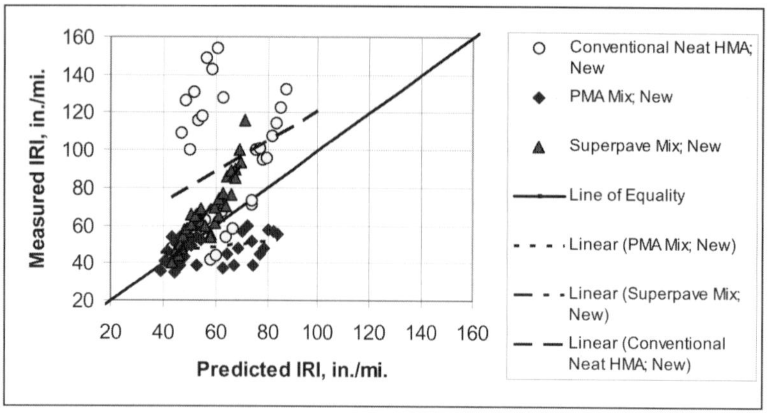

a. Intercept and slope estimators that are dependent on mixture type for the new construction PMS segments.

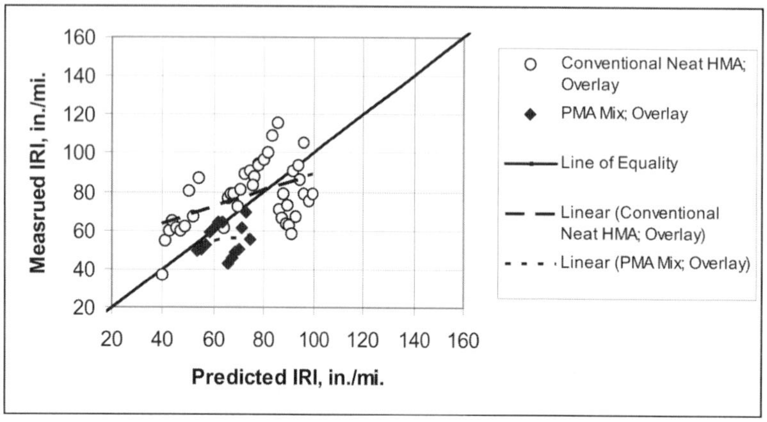

b. Intercept and slope estimators that are dependent on mixture type for the rehabilitation PMS segments.

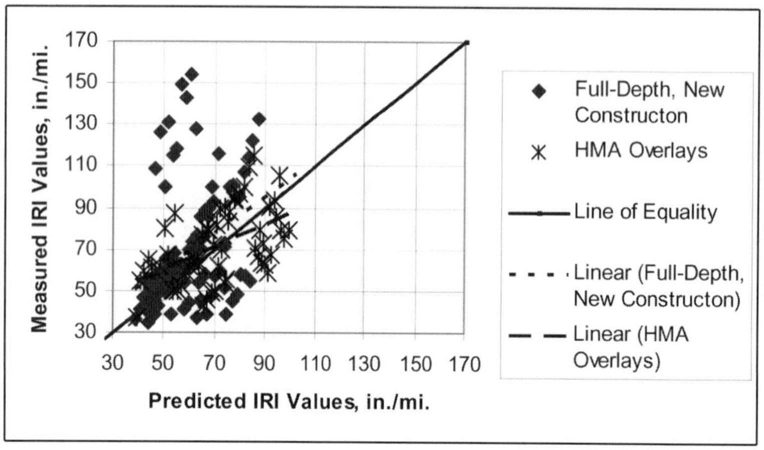

c. Intercept and slope estimators that are structure dependent for the PMS segments.

Figure A2-7. Comparison of the Intercept and Slope Estimators to the Line of Equality for the Predicted and Measured IRI Using the Global Calibration Values

Thus, the MEPDG global calibration values resulted in bias for rutting, transverse cracking, and IRI. The bias for fatigue cracking is low relative to the tolerable bias (refer to Table A2-6), because the measured and predicted values are low. The hypothesis, however, was rejected for all distresses at a 90-percent confidence level based on the slope and/or interceptor estimators. The following summarizes the findings for each transfer function or performance indicator.

- **Rut Depth**
  - Extensive dispersion exists between the predicted and measured rut depths within both sets of data—full-depth new construction and HMA overlays. Poor correlation exists between the predicted and measured rut depths (refer to Figure A2-2 and Attachment A2.4.B).
  - The MEPDGG over predicted the measured rut depths for the HMA overlays (refer to Figure A2-2). More importantly, the residual error is dependent on the pavement structure and type of mixture (refer to Figures A2-2 and A2-6).
  - The maximum rutting predicted in the subgrade and unbound layers varied from 0.1 to 0.3 in. A value of 0.3 in. already exceeds the measured rut depth for most of these PMS segments recorded in the database. In addition, previous forensic studies completed in Kansas on full-depth or depth strength projects with and without stabilized layers have suggested that the rutting in the unbound soils and materials is nil. Thus, it is hypothesized that the subgrade and unbound layer rut depths are over predicted. For this demonstration, it was assumed that the maximum rutting in the subgrade under thick HMA layers would be limited to a value of 0.1 in. in determining the local calibration value for these roadway segments. That local calibration value was then used to predict unbound layer rutting for the PMS segments with thinner HMA layers. This assumption is considered acceptable under the condition that all unbound layers were constructed in accordance with the project specifications.
  - The slope of the rut depth versus time or pavement age is lower for the HMA overlays than for the new construction full-depth segments. The truck traffic (Average Annual Daily Truck Traffic [AADTT]) is low for all of the PMS segments, with AADTT values being less than 100 for nearly half of the segments. Low truck traffic results in minimal increases in rutting over time, after the first couple of years.

- **Total Fatigue (Alligator) Cracking**
  - There are too few PMS segments with measurable fatigue cracking to validate or confirm the global calibration values and determine the local calibration values, if needed (refer to Figure A2-3). Twelve of the 16 PMS segments (75 percent) have none to less than 2 percent fatigue cracking over the monitoring period of time. The MEPDG consistently under predicted the measured fatigue cracks for those PMS segments exhibiting fatigue cracks.

- **Thermal (Transverse) Cracking**
  - Thermal cracking was predicted for only two (87 percent) of the full-depth PMS segments and those segments did not exhibit any transverse cracking, while large amounts of transverse cracks were recorded in the PMS database for many of the PMS segments of both data sets—full-depth, new construction and HMA overlays (refer to Figure A2-4).
  - A negative bias exists for the thermal cracking prediction model, even though the statistical analysis suggests that the bias is insignificant (refer to Table A2-7).

- **IRI or Roughness**
  - The IRI values are heavily dependent on the other distresses calculated by the MEPDG and the site factor. Changing the local calibration factor from unity will affect the IRI values. Thus, the IRI predictions should be evaluated for bias only after the bias has been removed from the other prediction models.

Step 8—Eliminate Local Bias of Distress Prediction Models
All of the globally calibrated transfer functions were found to be biased based on the intercept and slope estimators from Step 7. The process used to eliminate the bias depends on the cause of that bias and the accuracy desired by the agency. In general, there are three possibilities which are listed below.

1. The residual errors are, for the most part, always positive or negative with a low standard error of the estimate in comparison to the trigger value, and the slope of the residual errors versus predicted values is relatively constant and close to zero. The precision of the prediction model is reasonable but the accuracy is poor. In this case, the local calibration coefficient is used to reduce the bias. This condition generally requires the least level of effort and the fewest number of runs or iterations of the MEPDG to reduce the bias.

2. The bias is low and relatively constant with time or number of loading cycles, but the residual errors have a wide dispersion varying from positive to negative values. The accuracy of the prediction model is reasonable, but the precision is poor. In this case, the coefficient of the prediction equation is used to reduce the bias but the value of the local calibration coefficient is probably dependent on some site feature, material property, and/or design feature included in the sampling template. This condition generally requires more runs and a higher level of effort to reduce dispersion of the residual errors.

3. The residual errors versus the predicted values exhibit a significant and variable slope that is dependent on the predicted value. The precision of the prediction model is poor and the accuracy is time or number of loading cycles dependent—there is poor correlation between the predicted and measured values. This condition is the most difficult to evaluate because the exponent of the number of loading cycles needs to be considered. This condition also requires the highest level of effort and many more runs to reduce bias and dispersion.

The third one applies to this demonstration. An analysis of variance (ANOVA) was completed to determine whether e_r, b_o, and/or m are dependent on factors included in the sampling matrix or some other design feature and site condition factor of the PMS segments. As shown in Figures A2-2 and A2-5 the residual error is dependent on pavement structure and mixture type.

To eliminate the bias, the agency should first decide on whether to use the agency specific values or the local calibration factors that are considered as inputs in the MEPDG software. Either one can be used with success—for this demonstration the local calibration parameters were used. Time-history plots of each performance indicator should be prepared to determine if one or multiple calibration factors need to be evaluated, as noted above. The following describes the process used to eliminate the bias using the performance indicators stored in the Kansas PMS database.

Rut Depth Transfer Function
Poor correlation was found between the predicted and measured rut depths using the global calibration values, even though the bias is insignificant for both new construction and rehabilitation strategies (refer to Table A2-7 and Figure A2-2). One possible reason for the poor correlation is the large measurement error in rut depths. There is more variability in the measured rut depths within a PMS segment than between the segments. As an example, the standard deviation of the average maximum rut depth between all of the segments is 0.0754 in. (refer to Table A2-6), while the standard deviation of the average rut depths within the monitoring period for a specific PMS segment can be as high as 0.085 in. Varying the local calibration values to represent different site

conditions and materials will not increase model precision (i.e., reducing the data measurement errors). The following points or observations were identified in completing an analysis of the residual errors relative to the sampling template.

- The maximum predicted rutting in the unbound layers varied from 0.1 to about 0.3 in. This level of rutting in the unbound layers and foundation is inconsistent with previous forensic studies conducted by KSDOT. All PMS segments were constructed in accordance with KSDOT specifications. Thus, it is hypothesized that the rutting in the unbound layers is over predicted for these PMS segments.

The local calibration value for the unbound layers (β_{s1}) was estimated by making repeat runs of the MEPDG with varying values for a limited number of segments (four were used for this demonstration) to reduce the bias within each PMS segment. An average value of 0.50 was estimated for both the fine and coarse-grained soils (new construction and HMA overlays). An insufficient number of PMS segments with different soils were available to determine whether the local calibration value is soil type dependent. [Note: Version 1.0 of the MEPDG only allows the local calibration factor for the subgrade to be altered from unity; changes from unity cannot be made to the unbound aggregate base layers.]

- A review of the comparisons between the predicted and measured rut depths included in Attachment A2.4.B and in Figure A2-2 found that the MEPDG over predicted the measured rut depths for the HMA overlays, under predicted the rutting of the segments with Superpave and PMA mixtures, and over predicted the rutting of the conventional, neat HMA mixtures. Thus, the residual errors or bias on a sampling template basis is cell specific.

The local calibration exponents of temperature (β_{r2}) and number of load cycles (β_{r3}) in the rut depth transfer function were considered adequate, because of the variability in the measured rut depths with time—believed to be measurement error. However, the rut depth versus age relationship for some of the PMS segments has a greater slope than predicted with the MEPDG (see plots in Attachment A2.4.B). Thus, the local calibration parameter for the coefficient (β_{r1}) and number of load cycles (β_{r3}) were the terms considered in the local calibration process. The following summarizes the results from the ANOVA and local calibration process.

- For the full-depth HMA pavements, the residual error was found to be mixture dependent—conventional dense-graded versus polymer modified asphalt (PMA) versus Superpave coarse or gap-graded mixtures. However, correspondence or correlation was not identified between the residual error and volumetric properties of the HMA, layer thickness, and other input values. Thus, constant values for the HMA local calibration values were determined for each mixture type, which are listed below.
  - Conventional, Dense-Graded HMA Mixtures: $\beta_{r1} = 1.5$ and $\beta_{r3} = 0.9$
  - Superpave, Dense-Graded Mixtures: $\beta_{r1} = 1.5$ and $\beta_{r3} = 1.2$
  - Polymer Modified, Dense-Graded Mixtures: $\beta_{r1} = 2.5$ and $\beta_{r3} = 1.15$

- This mixture effect was not found for the HMA overlays, because all of the overlaid segments exhibited much lower rut depths. The average maximum rut depth measured along the new construction projects (full-depth segments) was 0.26 in., while an average maximum value of 0.175 in. was measured along the overlaid segments (refer to Attachment A2.4.B). The values determined for the two local calibration parameters for the HMA-Overlay Projects were: $\beta_{r1} = 1.5$ and $\beta_{r3} = 0.95$.

Table A2-8 lists the rut depth bias using the local calibration values listed above for the full-depth new construction and HMA-Overlay Projects. As shown, the hypothesis is accepted, and the statistical parameters indicate a more precise rut depth prediction model for the Kansas PMS segments (Table A2-7 compared to Table A2-8). Figure A2-8 compares the predicted and measured rut depth using the local calibration values, and shows increased precision, as compared to the use of the global calibration values (refer to Figure A2-2), especially for the new construction—full-depth segments. The rut depth transfer function still does not adequately explain the measured values for the HMA overlays.

Fatigue (Alligator) Cracking Transfer Function

Low- to no-fatigue cracking was predicted for the PMS segments, which exhibited little to no fatigue cracks. Thus, the bias is low. For those limited PMS segments with fatigue cracking, the local calibration coefficient (β_{f1}) was used to reduce the bias to the minimum value possible for those segments and that value was used to predict the fatigue cracking for all PMS segments.

The resulting local calibration value was less than 0.00005 for the HMA-Overlay Projects. This value is low and would result in much greater amounts of fatigue cracking if used for the full-depth new construction projects. Additional runs were made with the MEDPG assuming that bond had been lost between the existing surface and HMA overlay. Using that assumption, the resulting local calibration value was similar to the value determined for the full-depth new construction projects for the same type of mixtures that had exhibited low levels of fatigue cracking (β_{f1} =0.05). Thus, the local calibration value was determined using the condition of zero bond between the existing surface and HMA overlay. The local calibration value was found to be mixture dependent for the full-depth pavements.

Table A2-8. Summary of the Statistical Parameters—Local Calibration Values Used for Predicting the Performance Indicators for the Kansas PMS Sections

Performance Indicator	Project	Bias	Standard Error	s_e/s_y	R^2	Hypothesis; H_o:	Comment
Rutting	New	+0.0249	0.00397	0.522	0.650	Accept; p = 0.316	Transfer function is adequate.
	Rehab.	+0.0278	0.0725	1.26	Poor		Transfer function does not explain variation in measured data.
Fatigue Cracking	New	+0.383	2.154	0.814	0.322	Accept; p = 0.204	Transfer function does not explain variation in measured data.
	Rehab	+1.272	1.441	0.937	Poor		
Transverse Cracking	New	−59.4	313.6	0.626	0.595	Accept; p = 0.453	Transfer function is adequate.
	Rehab	−43.7	410.2	0.610	0.736		
IRI	New	−3.99	9.8	0.348	0.703	Accept; p =0.444	Regression model is adequate.
	Rehab	+0.38	14.3	0.864	0.402		
Rutting	SPS-1	+0.0059	0.209	0.877	0.252	Accept	Significantly under predicting total rutting; slope estimator significantly different than 1.0.
	SPS-5	−0.081	0.0989	1.46	Poor	Accept	

Table A2-8—Continued

Performance Indicator	Project	Bias	Standard Error	s_e/s_y	R^2	Hypothesis; H_o:	Comment
Fatigue Cracking	SPS-1	−8.63	17.34	1.05	Poor	Accept	Significantly under predicting fatigue cracking; slope and intercept estimators significantly different than line of equality
	SPS-5	−2.76	8.57	0.720	Poor	Accept	
Transverse Cracking	SPS-1	−173.2	408.3	0.857	Poor	Accept	Significantly under predicting transverse cracking; slope and intercept estimators significantly different than line of equality
	SPS-5	−638.2	1017.3	1.12	Poor	Accept	
IRI	SPS-1	−6.44	19.43	0.790	0.350	Accept	The hypothesis should be checked and evaluated after any bias has been reduced for the other distresses.
	SPS-5	1.08	8.507	0.973	0.372	Accept	

Note 1: Poor means that the model did not explain variation in the measured data within and between the LTPP test sections.
Residual Error = $e_r = y - x_i$
y_i = Measured or Observed Value; S_y = Standard Deviation of the observed values.
x_i = Predicted Value

The local calibration exponents of tensile strain (β_{f2}) and dynamic modulus (β_{f3}) in the allowable number of load applications equation were considered adequate because of the variability in the measured fatigue cracking with time—believed to be measurement error. The C_2 parameter in the bottom-up fatigue cracking prediction equation was also excluded from the evaluation. The C_2 value of unity determined from the global calibration effort was assumed to be appropriate for the Kansas PMS segments.

That assumption for C_2, however, is probably incorrect. The growth in fatigue cracking with time can be much steeper than predicted by the MEPDG using the global calibration values. This condition is illustrated in Figure A2-9 for the two PMS segments with higher amounts of fatigue cracking. This difference between the measured and predicted values with time will decrease the precision of the fatigue cracking prediction model for the Kansas PMS segments. Unfortunately, there are too few PMS segments with appreciable fatigue cracking to determine a reliable estimate of C_2. In addition, the measurement error and combining longitudinal cracks in the wheel path with the area fatigue cracks (alligator cracks) make it difficult to reliably estimate both C_2 and β_{f1}. The following summarizes the results from the local calibration process.

- For the full-depth HMA pavement (new construction), the HMA local calibration coefficient was found to be mixture dependent—conventional dense-graded versus PMA versus Superpave coarse or gap-graded mixtures. A constant value was determined for each mixture type, which are listed below.

- Conventional, Dense-Graded HMA Mixtures: $\beta_{f1} = 0.05$
- Polymer Modified, Dense-Graded Mixtures: $\beta_{f1} = 0.005$
- Superpave, Dense-Graded Mixtures: $\beta_{f1} = 0.0005$

- This mixture effect was not found for the HMA overlays, similar to the finding for rutting. The value determined for the local calibration parameter for the overlay projects is: $\beta_{f1} = 0.05$. One reason hypothesized for this finding is that inadequate bond exists between the existing surface and HMA overlay. However, other construction/material anomalies (i.e., segregation, stripping, etc.) may explain this difference. Forensic investigations would need to be completed to confirm or reject the inadequate bond hypothesis.

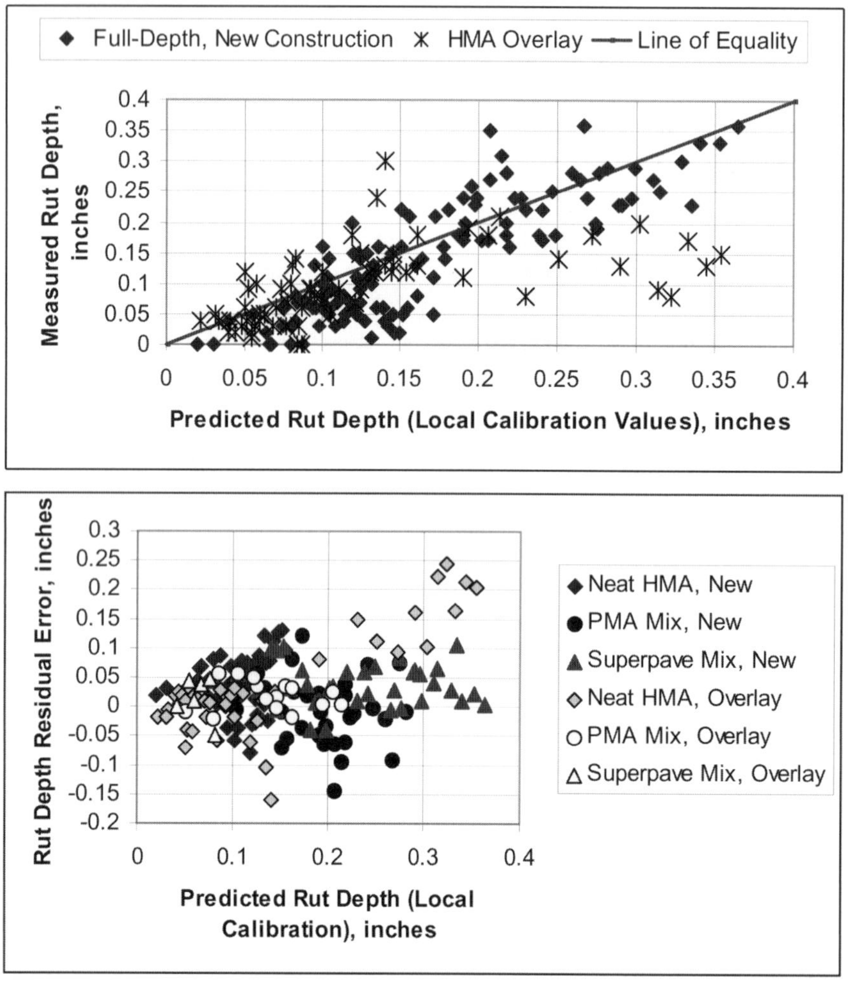

Figure A2-8. Comparison of Predicted and Measured Rut Depths Using the Subgrade and HMA Local Calibration Values for the PMS Segments

Table A2-8 lists the fatigue cracking bias using the local calibration values listed above. As shown, the hypothesis is accepted, but the statistical parameters still indicate a poor correlation between the predicted and measured values (Table A2-7 compared to Table A2-8). Figure A2-10 compares the predicted and measured fatigue cracking using the local calibration values, and shows that there is increased accuracy of the transfer function, as compared to the use of the global calibration values (refer to Figure A2-3).

Although the hypothesis for the local calibration values was accepted, these values would not be recommended for use from a practical engineering standpoint without more segments exhibiting fatigue cracking approaching the trigger value or design criteria. As noted above, 75 percent of the PMS segments have yet to exhibit measurable magnitudes of fatigue cracking and those with measurable cracking were significantly less than 10 percent, with the exception for one PMS segment (refer to Attachment A2.4.B).

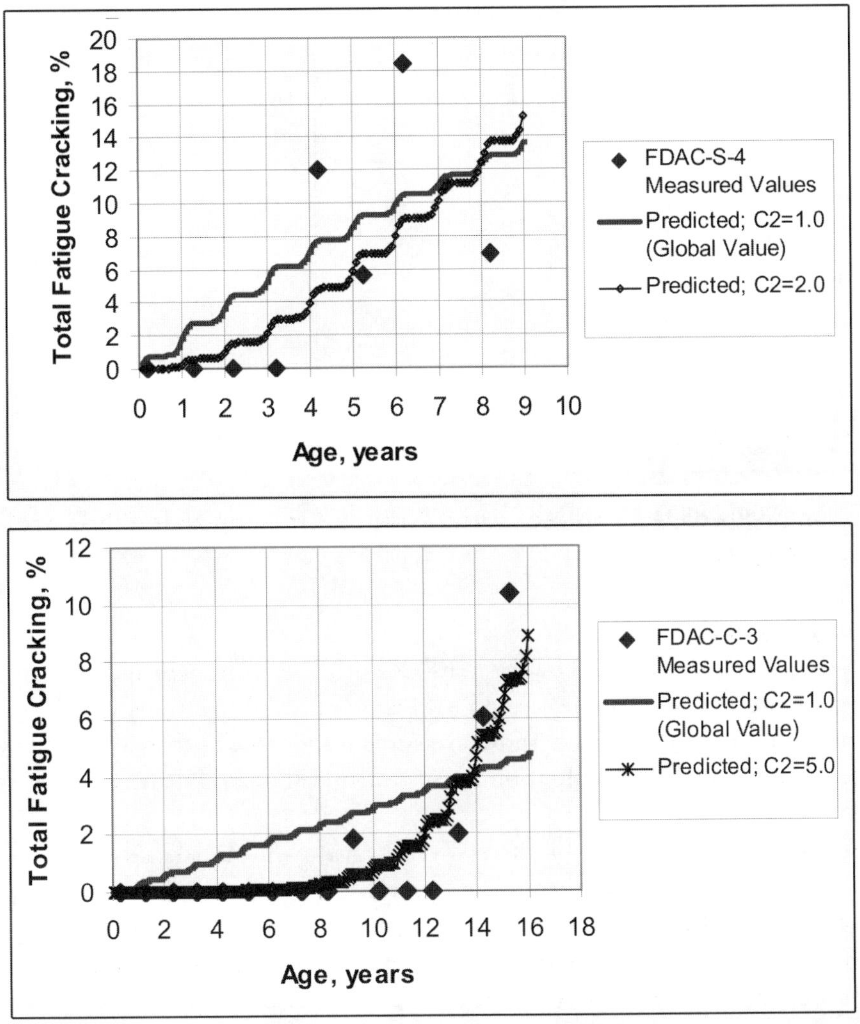

Figure A2-9. Comparison of Measured and Predicted Values of Fatigue Cracking Using Different Value for the C_2 and β_{f1} Parameters for PMS Segments FDAC-C-3 and FDAC-S-4

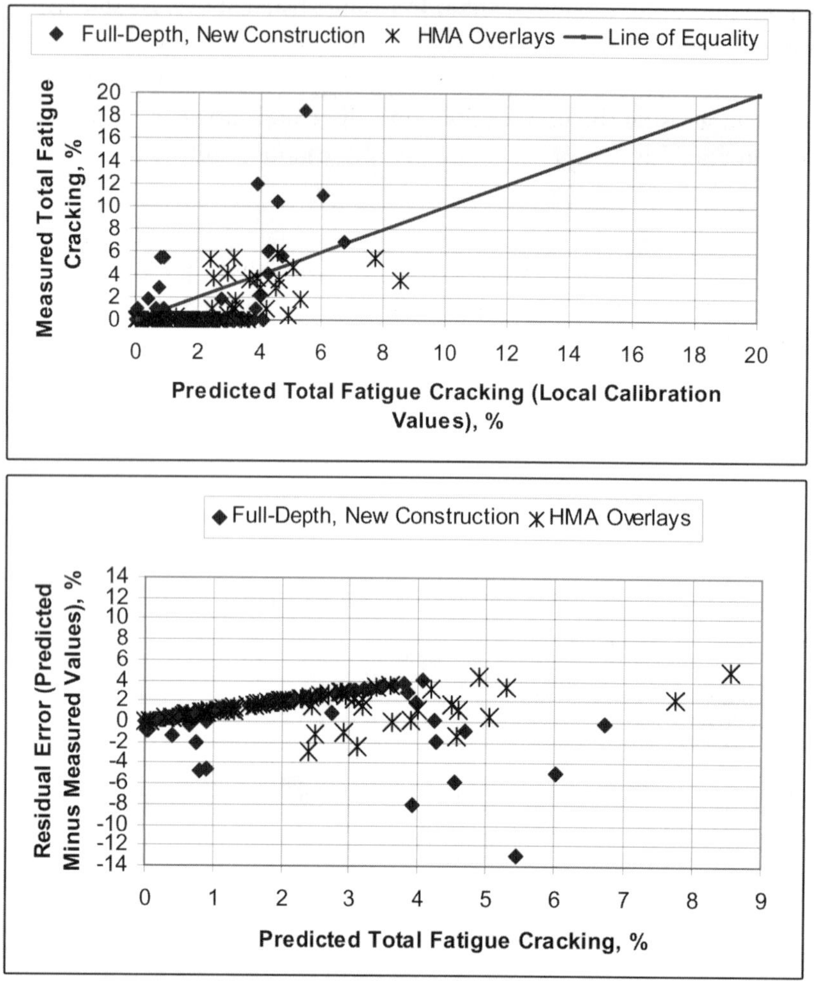

Figure A2-10. Comparison of Predicted and Measured Fatigue Cracking Using the Local Calibration Values for the PMS Segments

Transverse Cracking Transfer Function

The maximum length of thermal cracks predicted by the MEPDG is 2,200 ft/mi, which corresponds to about a 30-ft spacing of transverse cracks. Some of the PMS segments exhibited transverse cracking exceeding that maximum limit. Thus, only those measured responses less than about 2,500 ft/mi should be used in determining the local calibration factor to reduce bias and dispersion.

Figure A2-4 compared the observed and predicted transverse cracks using the global calibration values. As shown, the length of transverse cracks was under predicted for nearly all of the PMA segments for new construction and HMA overlays. In fact, thermal cracking was predicted for only two of the full-depth projects and none for the HMA-Overlay Projects. The local calibration parameter (β_{t3}) was used to reduce that bias (refer to Table A2-7).

In summary, the residual error was found to be mixture and structure dependent—conventional dense-graded versus PMA versus Superpave coarse or gap-graded mixtures; and new construction versus HMA overlays. The thermal cracking local calibration values (β_{t3}) to reduce model bias are listed below for the different mixtures and structures as shown in Table A2-9 (as defined by the sampling template [refer to Table A2-4]).

Table A2-9. Thermal Cracking Local Calibration Values

Mixture Type	Pavement Type	
	New Construction	HMA Overlay
Polymer Modified, Dense-Graded Mixtures	2.0	2.0
Conventional, Dense-Graded HMA Mixtures	2.0	7.5
Superpave, Dense-Graded Mixtures	3.5	7.5

Table A2-8 lists the thermal cracking bias using the local calibration values listed above. As shown, the hypothesis is accepted, and the statistical parameters indicate a more accurate and precise thermal cracking prediction model for the Kansas PMS segments (Table A2-7 compared to Table A2-8). Figure A2-11 compares the predicted and measured thermal cracking using the local calibration value, and shows that there is an increase in model accuracy and precision, as compared to the use of the global calibration values (refer to Figure A2-4). However, the null hypothesis for the intercept and slope estimators for the HMA overlays would still be rejected for the overlaid segments (refer to Figure A2-11).

Roughness or IRI Regression Model

The IRI values predicted by the MEPDG using the local calibration values for the other distresses are within acceptable limits of the measured values. Figure A2-12 compares the measured and predicted IRI values, while Table A2-8 summarizes the statistical information. As shown, the hypothesis was accepted—the IRI regression prediction equation is confirmed for the Kansas PMS segments.

If the hypothesis has been rejected, however, the agency would first identify the distress causing the higher residual errors or if the residual error is heavily time dependent (time and site factor related). The coefficients of the distress and/or site factor terms included in the IRI prediction equation would be determined to reduce local bias.

Step 9—Assess Standard Error of the Estimate

After the bias is reduced or eliminated for each of the transfer functions, the standard error of the estimate (refer to Table A2-8) is evaluated. The Standard Error of the Estimate (SEE) for each globally calibrated transfer function is included under the "Tools" section of the MEPDG software. Figure A2-13 compares the SEE for the globally calibrated transfer functions to the SEE for the locally calibrated transfer functions. For the runs using the local calibration values, the SEE was found to be statistically different in comparison to the SEE included in the MEPDG for each performance indicator. The following summarizes the comparison of the values between the global and local calibration.

- **Rut Depth Transfer Function (Total Rut Depths)**—The SEE values are lower for the locally calibrated transfer function than for the globally calibrated model. The PMS segments, however, exhibited much smaller rut depths than for the global calibration database.

- **Alligator Cracking Transfer Function**—SEE values based on the local calibration are lower than the values determined from the global calibration process. The PMS segments, however, exhibited less than 5 percent fatigue cracks for most of the test sections. This amount of cracking is too low in comparison to the trigger value (refer to Table A2-5) to develop a valid relationship between the SEE and predicted fatigue cracking values.

- **Thermal Cracking Transfer Function**—SEE values based on the local calibration are consistently higher than the values determined from the global calibration process.
- **IRI Regression Model**—SEE values for the IRI regression equation are not included on the MEPDG software screens and can not be changed.

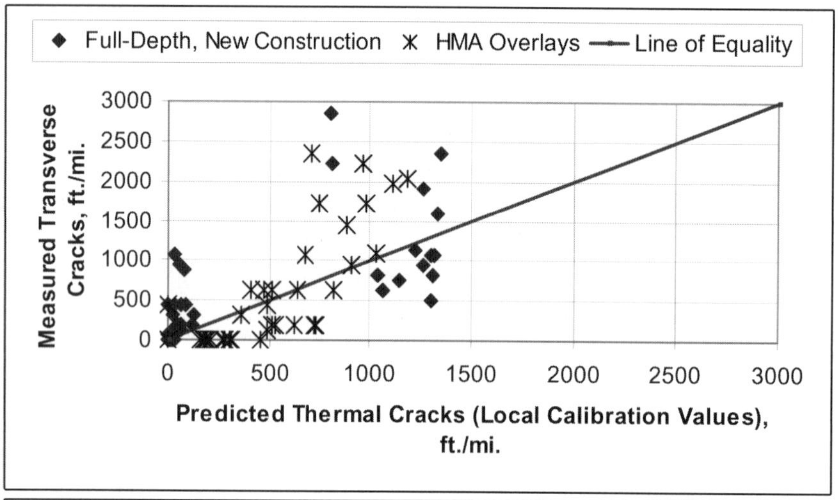

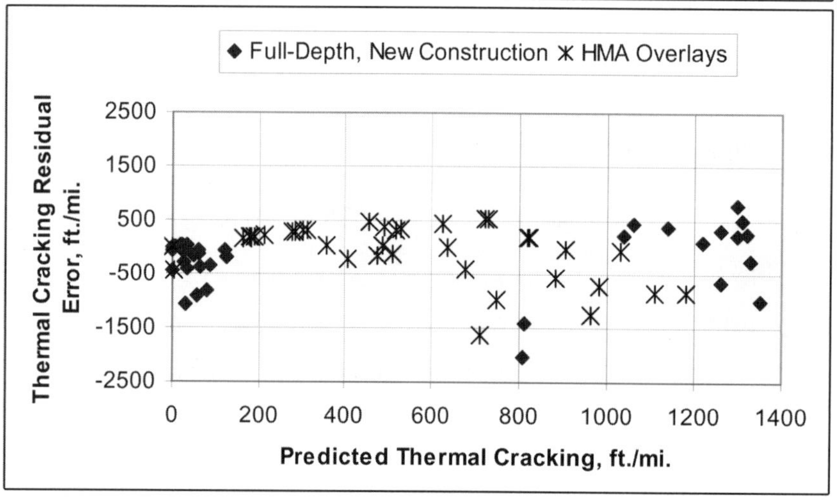

Figure A2-11. Comparison of Predicted Thermal Cracking and Measured Transverse Cracking Using the Local Calibration Value for the PMS Segments

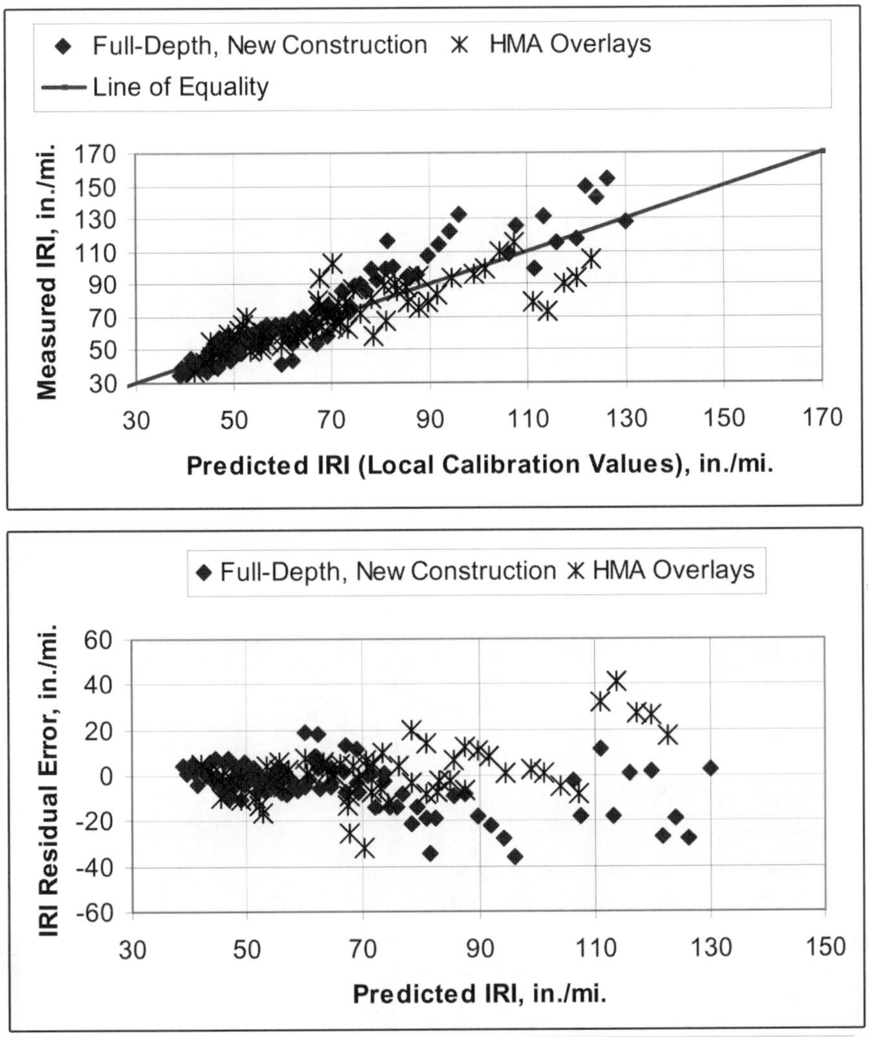

Figure A2-12. Comparison of Predicted and Measured IRI Values Using the Global Calibration Values

Step 10—Reduce Standard Error of the Estimate

As noted in Step 9 and shown in Figure A2-13, the SEE from the local calibration process was found to be different than the SEE relationships included in the MEPDG software for rutting, fatigue cracking, and thermal cracking. An ANOVA can be completed to determine if the residual error is dependent on some other parameter or material/layer property for the PMS segments. No correlation was identified, so the SEE values shown in Figure A2-13 and the local calibration factors summarized in Step 8 are believed to be the final values for the PMS segments included in the sampling matrix.

Step 11—Interpretation of Results and Deciding on Adequacy of Calibration Factors

For this demonstration, the global calibration values did result in a bias for all distresses. The MEPDG did not accurately explain the differences in performance between the different HMA mixtures and pavement structures. To reduce that bias required local calibration values that were different from unity. The MEPDG IRI regression equation was the only model that was confirmed using the KSDOT PMS data. The purpose of this step is to decide whether to adopt the local calibration values or continue to use the global values that were based on data included in the LTPP program from around the United States.

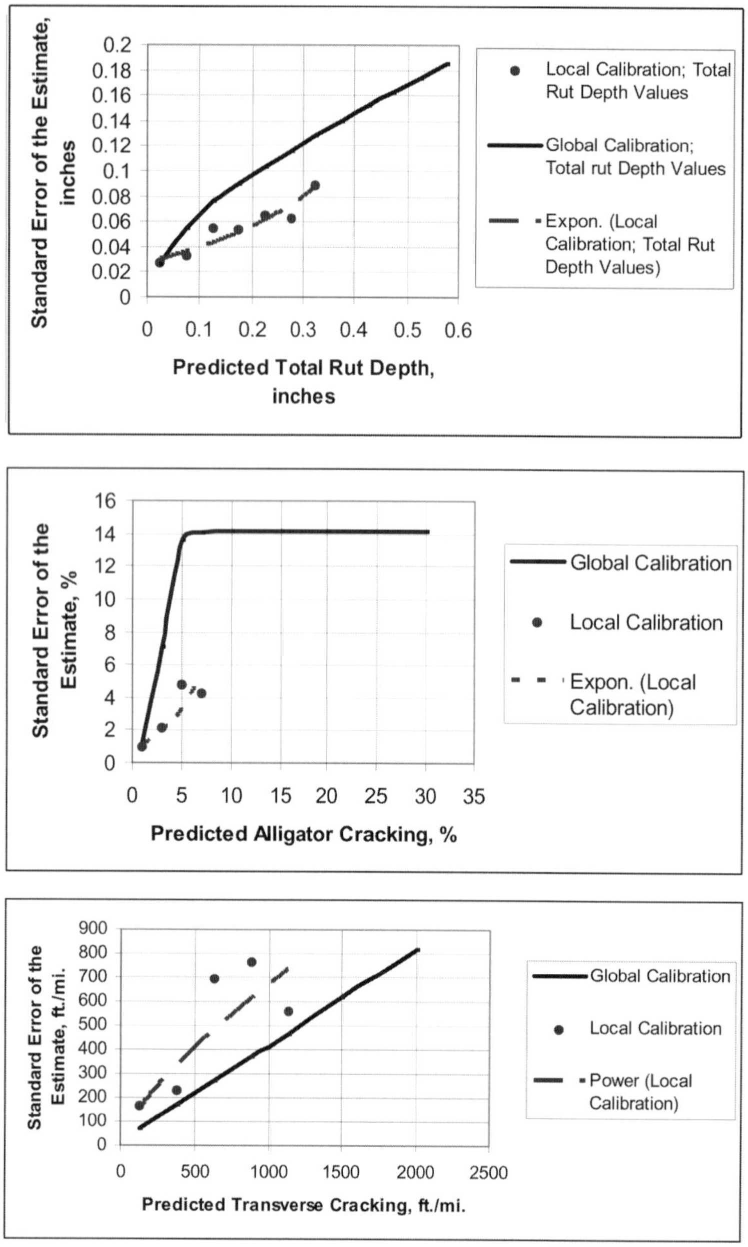

Figure A2-13. Comparison of the Standard Error of the Estimate for the Global-Calibrated and Local-Calibrated Transfer Function

To make that decision, an agency should identify major differences between the LTPP projects and the standard practice of the agency to specify, construct, and maintain their roadway network. Section 3.5 of the *MEPDG Manual of Practice* (*Design Features and Factors Not Included Within the MEPDG Process*) lists the factors and/or features that were excluded from the global calibration process. More importantly, the agency should determine whether the local calibration values can explain those differences. The agency should evaluate any change from unity for the local calibration parameters to ensure that the change provides engineering reasonableness.

The interpretation of results is discussed further in Section A2.3 (Summary for Local/Regional Calibration Values) using the two different data sets: PMS segments and selected LTPP SPS projects in and adjacent to Kansas. The following briefly interprets the results using PMS data.

- The IRI regression equation was found to be a reasonable simulation of the IRI values measured on the PMS segments in Kansas. Thus, the MEPDG IRI regression model is believed to be adequate for Kansas climate, materials, and other site features for their more common design strategies and mixtures.

- All HMA mixtures included in the PMS segments exhibited less resistance to fracture or are much more susceptible to fracture than included in the global calibration process. These mixtures are brittle in comparison to those used to determine the global calibration values. Although only small amounts of fatigue cracking have occurred, the HMA layers are thick and the truck traffic low. It is expected that many of these sections may have exhibited surface initiated cracking or the cracking recorded as fatigue in the Kansas PMS database is some other type of cracking caused by a combination of environmental conditions and wheel loads. Since the amount of cracking is low, the fatigue cracking local calibration values should not be used without additional sections exhibiting higher amounts of fatigue cracks. The C_2 parameter seems to be significantly different from unity (refer to Figure A2-9). However, the global calibration value for C_2 (unity) should continue to be used until more segments with higher amounts of fatigue cracking become available to confirm or dispute that observation. There are an insufficient number of PMS segments with higher levels of fatigue cracking to confirm the SEE at the trigger value (design criteria).

- All mixtures are also more susceptible to thermal cracking than those included in the global calibration process, similar to the finding for fatigue cracking. Substantial lengths of transverse cracking were exhibited on many of the PMS segments. Thus, it would be recommended that the local calibration value for thermal cracking be used for design. It would also be recommended that the SEE values determined from the local calibration process be used for design (refer to Figure A2-13).

- The Superpave and PMA mixtures exhibit more rutting potential than the conventional neat HMA mixtures used in Kansas. In summary, the neat HMA mixtures exhibit lower rutting potential than the mixtures included in the global calibration database, while the Superpave and PMA mixtures exhibit slightly higher levels of rutting. However, the magnitude of rutting is low and would not trigger any type of rehabilitation. The HMA local calibration values for rutting would be recommended for use. The SEE values derived from the local calibration are lower than the SEE values derived from the global calibration, but the measured rut depths for the PMS segments are significantly lower than the trigger value (refer to Table A2-5). There are an insufficient number of PMS segments with greater rut depths to confirm the SEE at the trigger value (design criteria).

- The subgrade rutting local calibration value is believed to be reasonable because of the findings from previous forensic studies and would be recommended for use.

In summary, other results or observations from the local validation-calibration process are listed below.

- The Superpave designated mixtures are inferior to the other mixtures used for new construction projects. These Superpave mixtures are more susceptible to fatigue cracking, rutting, and thermal cracking. The conventional dense-graded mixtures have the overall better performance characteristics for new construction projects.

- Mixture performance differences were not present for the HMA-Overlay Projects, with the exception for thermal cracking. The use of zero bond or interface friction probably minimizes the effect of mixture type differences on the local calibration values and performance, as noted in the following observation.

- The HMA overlays have exhibited higher levels of cracking than expected. For the same fatigue cracking local calibration value (assuming that the truck traffic levels are correct), the interface friction or bond between the HMA overlay and existing surface would need to be zero. As stated above, the global calibration values for fatigue cracking should be used until more roadway segments are included with higher amounts of fatigue cracking. The condition of bond between the existing surface and HMA overlay should be evaluated during design to ensure that the overlay thickness is adequate for both conditions—with and without bond.

A2.2 Demonstration 2—LTPP Data and Local Calibration

KSDOT had an insufficient number of research grade and LTPP test sections that could be identified within a reasonable time frame to execute the MEPDG. Thus, LTPP test sections within Kansas and some of the nearby states were used to demonstrate the local validation-calibration process. The LTPP SPS-1 and SPS-5 experiments included in the demonstration are listed below in Table A2-10.

Table A2-10. LTPP SPS-1 and SPS-5 Site Locations

SPS-1 Experiment		SPS-5 Experiment	
State	**Number of Sections**	**State**	**Number of Sections**
Kansas	6	Colorado	8
Nebraska	6	Missouri	8
Oklahoma	6	Oklahoma	8
Iowa	6	Texas	8

A2.2.1 Description of LTPP Test Sections Used in Demonstration

The test sections selected for use from the LTPP experiments were selected to be somewhat consistent with the type of pavements and materials used for the PMS demonstration. For the SPS-1 projects, the test sections included in the local validation-calibration process excluded those with asphalt treated permeable base (ATPB) layers. ATPB layers are not considered a typical design feature in Kansas, and none of the PMS segments included ATPB layers. For the SPS-5 projects, all test sections with and without recycled asphalt pavement (RAP) were included in the local calibration process, excluding the control section. Attachment A2.4.C describes the LTPP projects used in this demonstration and some of the construction problems reported for those projects. The construction problems reported may help explain potential anomalies identified during the validation-calibration process.

A2.2.2 Step-by-Step Procedure

Step 1—Select Hierarchical Input Level

This demonstration using LTPP projects is for the condition for which detailed data are available to determine many of the input parameters. Input Levels 2 and 3 were used for most input parameters—most were input Level 2. The general information from which the inputs were determined for each input category is discussed in Step 5.

Step 2—Experimental Factorial and Matrix or Sampling Template

The same sampling template for the PMS demonstration was used for this example, with two exceptions. The pavement cross sections with thinner HMA layers and thicker unbound aggregate base layers were used (defined as conventional flexible pavements) but only one type of mixture (defined as conventional neat HMA) with and without recycled asphalt pavement (RAP). Table A2-11 shows the sampling template for this demonstration. As shown, this demonstration is for the condition with a balanced factorial (sampling template) with replication.

Table A2-11. Sampling Template for the Demonstration Using LTPP Data

HMA-Mixture Type	New Construction		HMA Overlays		Total LTPP Sections
	Conventional Flexible Pavements	Full-Depth and Deep-Strength	Mix Without RAP	Mix with RAP	
Conventional Neat HMA	8	16	16	16	56
Superpave Mix	—	—	—	—	—
PMA Mix	—	—	—	—	—

Step 3—Estimate Sample Size for Each Performance Indicator Prediction Model

The minimum sample size or number of roadway projects required for the PMS segments would also apply to this demonstration (refer to Table A2-5). The number of LTPP test sections used for this demonstration for the conventional neat HMA mixtures is considered adequate for the level of significance and tolerable bias selected (Table A2-5 compared to Table A2-11).

Step 4—Select Roadway Segments

All of the applicable test sections from the SPS experiments were used for this demonstration (excluding those with an ATPB layer). Fifty-six test sections were selected for this example using distress data consistent with the global calibration process. The 56 test sections are sufficient for the three pavement structures (conventional and deep-strength new construction and HMA overlays of flexible pavements). Different types of HMA mixtures were not considered within this demonstration, with the exception of those with and without RAP. The projects and test sections used in the demonstration represent a balanced experimental matrix, unlike the one used for the PMS segments.

Step 5—Extract and Evaluate Distress and Project Data

This step is grouped into four activities: (1) extracting and reviewing the performance data; (2) comparing the performance indicator magnitudes to the trigger values; (3) evaluating the distress data to identify anomalies or outliers; and (4) determining the inputs to the MEPDG.

Step 5.1—Extract, Review, and Convert Measured Data to the Values Predicted by the MEPDG, if Needed.
The LTPP distress or performance indicator measurements were used within this demonstration. Thus, they are consistent with the values predicted by the MEPDG and no adjustment is needed.

The average measured value should be determined for each measurement period for each LTPP test section. The time-history data should not be smoothed but can be cleaned to remove any data errors. As an example, measured distresses that are recorded as zero after multiple values have been recorded with higher levels of distress. Plots of the average time-history data for the LTPP sections are included in Attachment A2.4.D. Some of the important observations of the data that could affect the validation-calibration process are listed below.

- There is dispersion in the measured distress data within the LTPP SPS sections—similar to the PMS segments. However, none of the LTPP SPS projects exhibit abrupt changes in the distress values measured over time. Improving on the precision of the prediction models is still considered not likely.
- Only one SPS-5 project (Colorado) exhibited high levels of fatigue cracking.
- More of the SPS-5 projects have higher levels of transverse cracking than the SPS-1 projects.
- Only one SPS- project (Nebraska) consistently exhibited high levels of rutting. The only other sections that exhibited high levels of rutting were a couple sections within the Kansas SPS-1 project.

Step 5.2—Compare Distress Magnitudes to the Trigger Values
The next activity of this step is to compare the distress magnitudes to the trigger values for each distress (see Table A2-5). Then answer the question—Does the sampling template include values close to the trigger values or design criteria?

Table A2-12 summarizes the average, maximum, and minimum distress values for each performance indicator, as compared to the design or trigger values for major rehabilitation. As shown, the average maximum distress values are within one or two standard deviations for all performance measures. Table A2-12 also summarizes the expected probability that the trigger values will be exceeded using the LTPP local calibration data set. The standard deviations, average maximum values, and probability of exceeding the trigger values for rutting and fatigue cracking for the LTPP SPS sites are significantly greater than those for the PMS roadway segments. This difference was expected because of the experimental factor effects designed into the LTPP experiments, while the same design procedure and material specifications were used for all of the PMS roadway segments.

Table A2-12. Summary of the Maximum Values of Different Performance Indicators in Comparison to the Design Criteria or Trigger Values (Number of Sections = 56)

Distress or Performance Indicator	Design Criteria	Maximum Values Measured for Each Test Section				Probability of Exceeding Trigger Values, %
		Average Max. Value	Lowest Max. Value	Largest Max. Value	Stand. Dev. of Max. Values	
Rut Depth, in.	0.40	0.40	0.157	1.181	0.251	50.0
Fatigue Cracking, %	20	13.3	0.0	61.0	20.29	37.1
Transverse Cracks, ft/mi	1,500	610	0	4,085	882	15.5
Roughness, in./mi	130	83.6	31.3	184.7	25.1	3.3

Step 5.3 —Evaluate Distress Data to Identify Anomalies and/or Outliers

As for the PMS roadway segments, the LTPP SPS data were reviewed to determine if any data should be removed from the validation-calibration database. Attachment A2.4.D includes the time-history plots of the measured distresses for each test section. The same two questions included under Demonstration 1—Step 5.3 should be considered when reviewing the performance data (refer to Step 5.3 under Demonstration 1 for a discussion of the items included under each question). The following provides some examples relative to those two questions, while Attachments A2.4.C and A3.2.B provide an evaluation and comparison of the performance indicators between and within the LTPP projects.

- Does the data make sense? The performance data for all of the LTPP SPS projects seem reasonable with the following exception:
  - The Texas SPS-5 exhibited higher rutting and transverse cracking than any of the other HMA-Overlay Projects. The Texas sections also had the lowest air void levels reported during construction (in place air voids vary from 3 to 5 percent). High levels of rutting are typically associated with low levels of transverse cracking. However, this project had plant problems during production and produced the HMA mixtures at elevated temperatures. This SPS-5 project could represent an anomaly within the LTPP SPS projects included within this demonstration.

- Are there test sections with anomalies, outliers, or blunders in the database? Potential anomalies or outliers were identified in the LTPP SPS data for the projects included within this demonstration. Some are listed below.
  - Figure A2-13 shows an example of the rut depths measured with pavement age for the Kansas SPS-1 test sections. As shown, some sections exhibited rut depths that increase and then decrease over time. This variation in the measured response is considered measurement error and all points should remain in the database. More importantly, section 20-0102 exhibited much more rutting than any of the other sections. This section is believed to be an anomaly, because of construction difficulties and rain delays noted in the construction report for the unbound layers. The water content of the unbound layers for sections 20-0101 and 20-0102 was much greater than in the other sections. More importantly, a minor amount of fly ash was added to the granular base of some sections to facilitate construction (refer

to Attachment A2.4.C). This data should also remain in the database, because it does not represent an error or blunder in the data—but is considered an anomaly.

- The Nebraska SPS-1 project consistently exhibits higher levels of rutting than any of the other SPS-1 and SPS-5 projects. The only other test sections with rut depths exceeding those of the Nebraska project are a couple of sections in the Kansas SPS-1 project (refer to Figure A2-14).
- The Colorado SPS-5 project exhibits higher levels of fatigue cracking than any of the other SPS-5 projects.

Figure A2-14. Rut Depths Measured Over Time for the Kansas SPS-1 Project

Step 5.4—Inputs to the MEPDG for Each Input Category
The MEPDG *Manual of Practice* should be used to determine the inputs to the MEPDG. Procedures to determine the inputs should be established based on the policies of the user agency. Once established, those procedures should be followed to ensure consistency of designs within the agency.

The following provides a discussion on the information extracted from the LTPP database for determining the inputs needed to execute the MEPDG for each of the LTPP test sections.

- **Initial IRI**—The initial IRI was measured immediately after construction. In a few cases, the initial IRI was unavailable. For these LTPP projects the initial value was determined from the measured values within two years after construction. This value is believed to be reasonable, because only very minor magnitudes of distress were recorded for the first couple of years after construction.

- **Cross Sections and Layer Thicknesses**—Included in the LTPP database. The average thickness for each layer was used.

- **Rehabilitation Inputs**—Input Level 2 was used. The condition of the HMA pavement prior to overlay for the SPS-5 sections was determined from the distress values included in the LTPP database prior to rehabilitation. Trenches were unavailable for the SPS-5 projects. A rut depth of 0.1 in. was assumed for the unbound layers (refer to discussion on forensic studies under Steps 6 and 7 of Demonstration 1). All layers were assumed to be fully bonded.

- **Traffic**—All traffic input parameters were extracted from the LTPP traffic database or tables. When those values were not included in the LTPP traffic tables, default values were used, with the exception of speed. The posted speed limit was used as the input for all SPS projects. In addition, the historical and monitoring traffic data included in the LTPP database were used to estimate the average growth in truck traffic for project. Axle weight data were unavailable for some of the SPS projects included within this demonstration. The default axle weight distributions included in the MEPDG were used for this demonstration.

- **Climate**—The longitude and latitude of each SPS project was used to create a virtual weather station for that project.

- **Materials**—The material and layer properties for each pavement layer and subgrade were extracted from the LTPP database. Volumetric properties are only available for the thicker layers, but not for every test section within an LTPP SPS project. No materials tests were completed on the thin layers, as defined by LTPP. The average layer properties were used for those LTPP SPS test sections without any test results to measure the volumetric properties required as inputs to the MEPDG.

  – Some of the properties needed as inputs to the MEPDG are excluded from the LTPP database (for example, dynamic modulus and creep compliance of the HMA mixtures). For these properties, Level 3 inputs were used or the property was calculated from other data (for example, the effective asphalt content by volume).

  – If adequate data were unavailable or missing for a specific layer and test section, the average property value for that layer was used as the input from the test results of that layer for that LTPP SPS project. If testing was not completed on a specific layer or material on any test section within a specific project, default values were used as the input or the value was based on other data sources of similar materials.

  – Resilient modulus data were available for all soils and unbound aggregate base layers for most of the SPS-1 projects. Input Level 1 was not used in the demonstration because of the run time required and complexity. However, the laboratory resilient modulus values were used to estimate the input value for this demonstration. Figures A2-15 and A2-16 show the repeated load resilient modulus values measured in the laboratory (LTPP test protocol) for the subgrade soil and dense graded aggregate base included in the Kansas SPS-1 project, respectively.

    The following procedure was used in determining the resilient modulus for each project. Test section 20-0104 (full-depth HMA section) was used to demonstrate the iterative procedure for determining the in place resilient modulus of the soil, while test section 20-0102 (conventional flexible pavement section) was used for the dense graded aggregate base layer.

    1. Determine or estimate the modulus of all bound layers within the pavement structure. The MEPDG was used to estimate the dynamic modulus within different seasons for the HMA and Asphalt Treated Base (ATB) layers. Typical values for the winter and summer months were used because they represent the greatest range in values throughout the year.

2. Calculate the at-rest stresses in the subgrade soil and unbound aggregate base layer. The thicknesses and densities of each layer were extracted from the LTPP database. The at-rest stress condition was calculated at the one-quarter depth into the unbound aggregate base layer and 18 in. into the subgrade layer.

3. Use WINJULEA, or another elastic layer program, to calculate the in place stresses within the unbound aggregate layers and subgrade for different seasons and axle loads at the same depths used to calculate the at-rest stress condition. A range of resilient modulus values were used for the subgrade and dense graded aggregate base layers. For the subgrade values from 6 to 14 ksi were used, while for the aggregate base layer the values ranged from 5 to 35 ksi.

4. Combine the computed stresses with the at-rest stresses at specific depths for estimating the laboratory resilient modulus that would be consistent with the elastic layer values.

5. Figure A2-15 (test section 20-0104, a full-depth structure) shows the total stress state computed in the subgrade layer for different seasons and axle loads. A light single axle load (12-kip) and dynamic modulus values for winter were used because this would result in much lower stresses in the subgrade, while a heavier single axle load (24-kip) in the warmest summer month was used to cover a diverse range in stress states. The in-place confining pressure or total lateral stress of 1.7 to 2 psi was determined from these computations. As shown, the resilient modulus for which the laboratory and theoretical values are equal varies from 9.9 to about 10.4 ksi. An overall average value of about 10 ksi was used as the input for each of the Kansas SPS-1 test sections.

6. Figure A2-16 (test section 20-0102, a conventional flexible structure) shows the stress state computed in the aggregate base layer for different seasons and axle loads, similar to Figure A2-12. As shown, the resilient modulus, at which the laboratory and theoretical values within the base layer are equal, vary from 12 to about 20 ksi. An overall average value of about 18 ksi was used as the input for the dense graded aggregate base layer for each of the SPS-1 test sections in Kansas.

Step 6—Conduct Field and Forensic Investigations

As noted for the PMS segments, KSDOT decided to accept the assumptions and conditions included in the MEPDG for the global calibration effort. No field and forensic investigations were conducted to determine the location of crack initiation and the amount of rutting within each pavement layer and foundation for the Kansas SPS-1 project. In addition, field forensics studies have yet to be completed for any of the other SPS projects included within this demonstration.

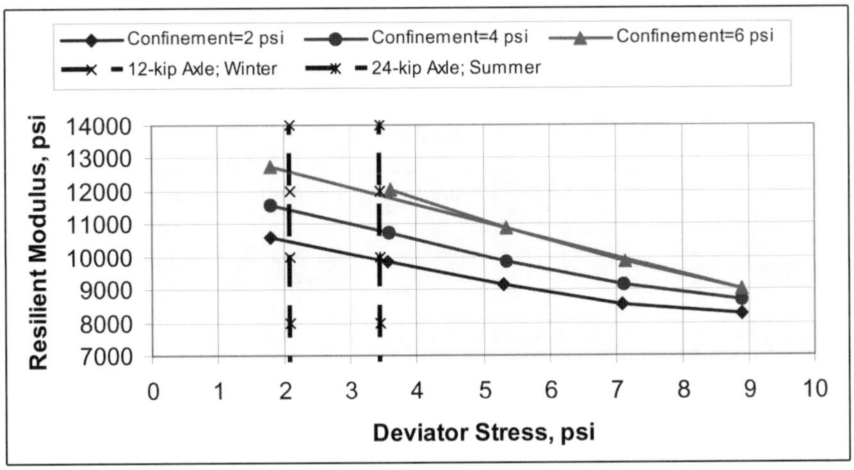

Figure A2-15. Determination of the Resilient Modulus of the Subgrade Using Laboratory Test Data Included in the LTPP Database

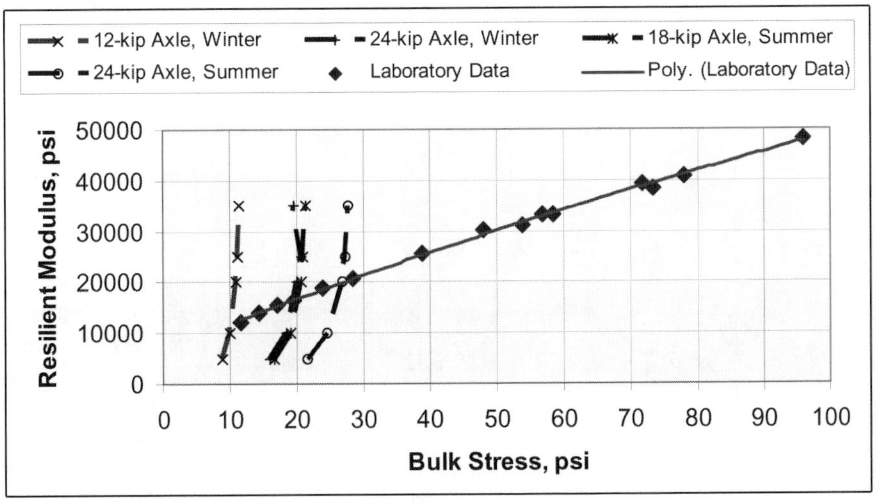

Figure A2-16. Determination of the Resilient Modulus of the Unbound Aggregate Base Layer Using Laboratory Test Data Included in the LTPP Database

Step 7—Assess Local Bias from Global Calibration Factors

The performance indicators are predicted with the MEPDG and compared to the measured values. Table A2-13 summarizes the statistical values for the SPS-1 (new construction) and SPS-5 (HMA overlays) projects, while Figures A2-17 through A2-20 compare the predicted and measured values for each performance indicator. The null hypothesis is that the average residual error for the two pavement structures (SPS-1 and SPS-5 projects) is zero at a specific confidence level (refer to Eq. A2-3 under Step 7 of Demonstration 1). A 90-percent confidence level was used within this demonstration, as for Demonstration 1 using the PMS segments.

Table A2-13. Summary of the Statistical Parameters—Global Calibration Values Used to Predict the Performance Indicators of the LTPP SPS-1 and SPS-5 Projects

Performance Indicator	Project	Bias, e_r	Standard Error, s_e	s_e/s_y	R^2 (See Note 1)	Hypothesis; H_o: $y_i - x_i = 0$	Comment
Rutting	SPS–1	+0.0059	0.209	0.877	0.252	Accept	Significantly under predicting total rutting; slope estimator significantly different than 1.0.
	SPS–5	–0.081	0.0989	1.46	Poor	Accept	
Fatigue Cracking	SPS-1	–8.63	17.34	1.05	Poor	Accept	Significantly under predicting fatigue cracking; slope and intercept estimators significantly different than line of equality.
	SPS-5	–2.76	8.57	0.720	Poor	Accept	
Transverse Cracking	SPS-1	–173.2	408.3	0.857	Poor	Accept	Significantly under predicting transverse cracking; slope and intercept estimators significantly different than line of equality.
	SPS-5	–638.2	1017.3	1.12	Poor	Accept	
IRI	SPS-1	–6.44	19.43	0.790	0.350	Accept	The hypothesis should be checked and evaluated after any bias has been reduced for the other distresses.
	SPS–5	1.08	8.507	0.973	0.372	Accept	

Note 1: Poor means that the model did not explain variation in the measured data within and between the LTPP test sections.
Residual Error = $e_r = y_i - x_i$
y_i = Measured or Observed Value; S_y = Standard Deviation of the observed values.
x_i = Predicted Value

Results from the hypothesis testing are summarized in Table A2-13. As shown, the hypothesis would be accepted for all performance indicators. The reason that the hypothesis was accepted is that the bias is low in comparison to the standard error and the within test section variability of the measured values.

The slope and intercept estimators were also used to evaluate model bias as for the PMS segments (refer to Step 7 under Demonstration 1). The slope and intercept estimators are cell dependent, as shown in Figures A2-17 through A2-20. In summary, all transfer functions exhibited similar trends or bias and that bias is related to pavement structure. Thus, the MEPDG global calibration values resulted in bias for each transfer function, similar to the results for Demonstration 1 using the PMS segments. Table A2-14 provides a brief description on the predictions of distress for the LTPP projects. The following summarizes the findings for each performance indicator.

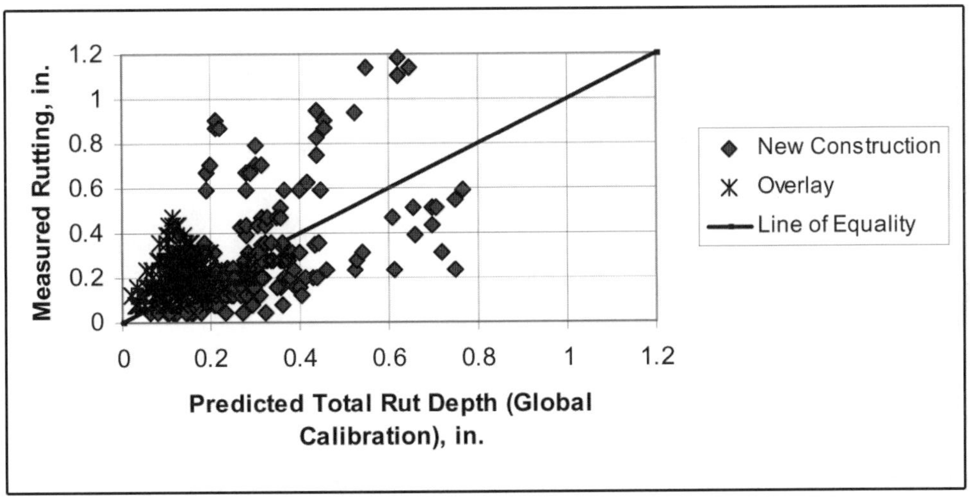

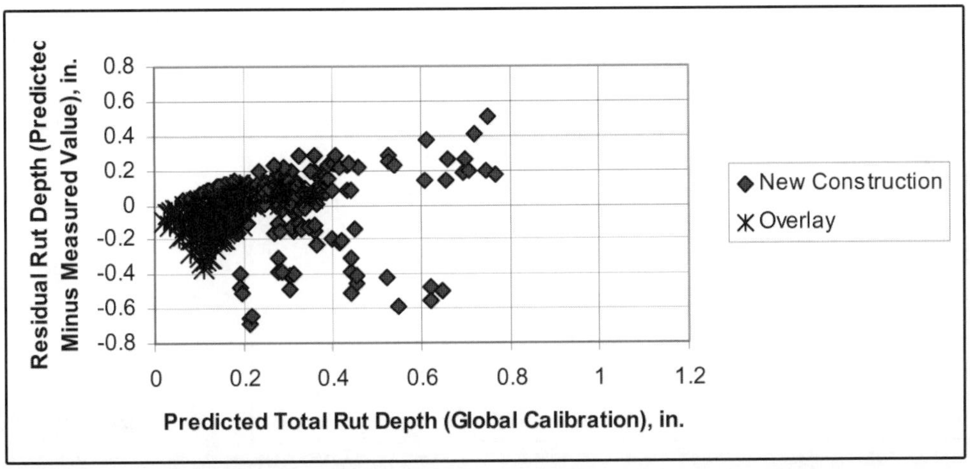

Figure A2-17. Comparison of Predicted and Measured Rut Depths Using the Global Calibration Values

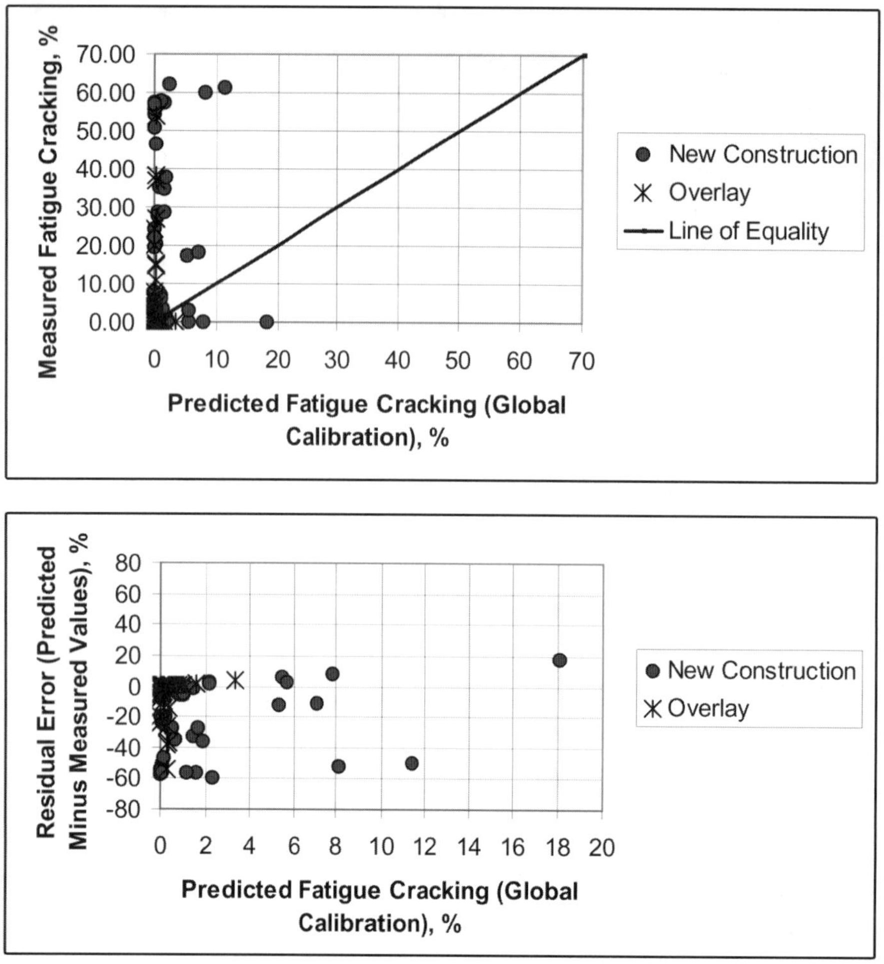

Figure A2-18. Comparison of Predicted and Measured Fatigue Cracking Using the Global Calibration Values

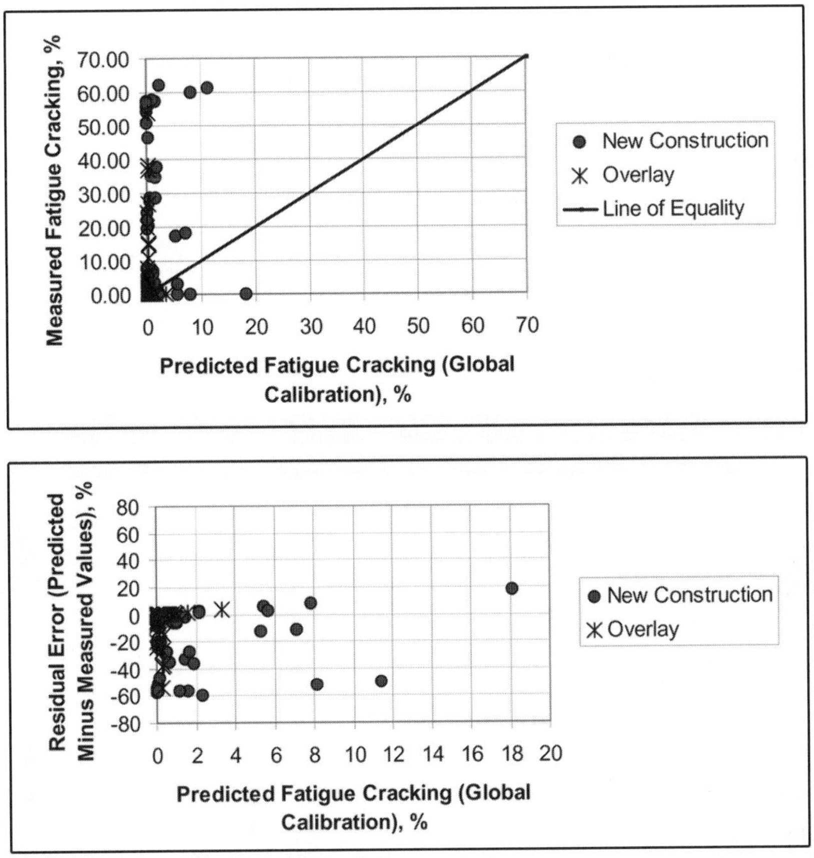

Figure A2-19. Comparison of Predicted Thermal Cracking and Measured Transverse Cracking Using the Global Calibration Values

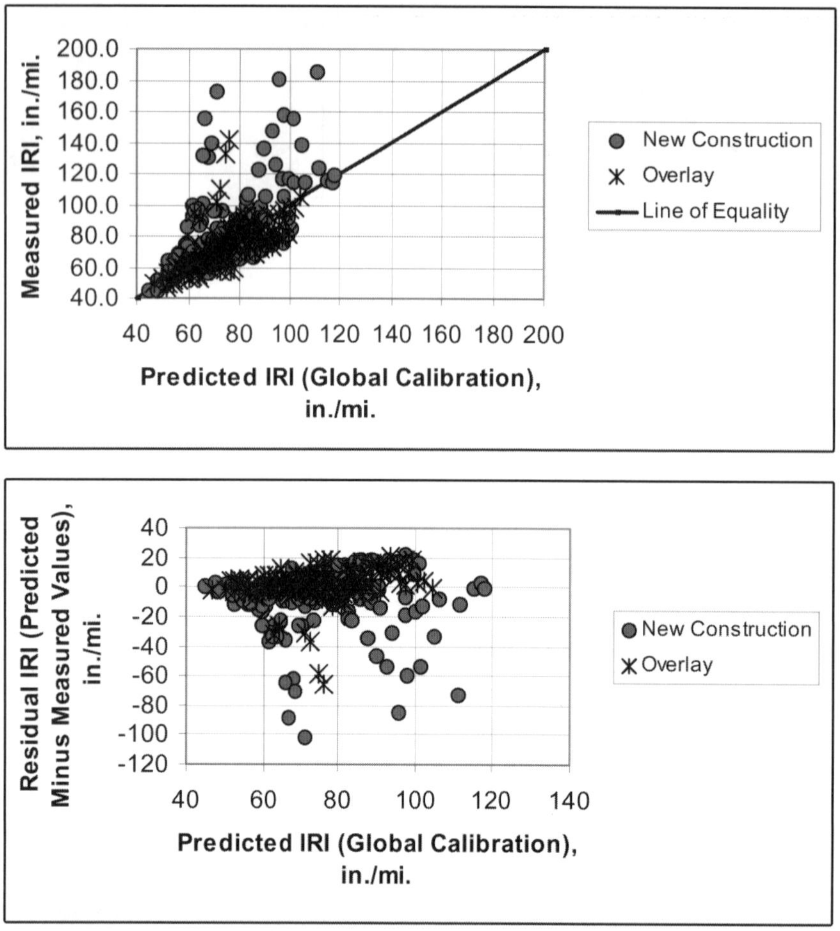

Figure A2-20. Comparison of Predicted and Measured IRI Using the Global Calibration Values

Table A2-14. Summary of the Performance Indicator Predictions Using the Global Calibration Values

Distress or Performance Indicator	New Construction (SPS-1 Projects)		HMA Overlay (SPS-5 Projects)	
Total Rut Depth	Kansas	Minimal bias, with the exception for 20-0102, under predicted rutting.	Colorado	Significant positive bias; over predicted rutting.
Total Rut Depth	Iowa	Significant positive bias; over predicting rutting. Sections 19-0102 & 19-0105 have a much higher positive bias.	Missouri	Significant negative bias; under predicted rutting.
Total Rut Depth	Oklahoma	Variable residual errors; Sections 40-0115, 40-0116, and 40-0117 have a minimal negative bias; while 40-0113, 40-0114, and 40-0118 have a significant positive bias.	Oklahoma	Minimal bias; but sections 40-0502 and 40-0508 (virgin mix sections) have a higher positive bias.
Total Rut Depth	Nebraska	Significant increase in measured rutting over time not predicted by MEPDG, resulting in a significant negative bias.	Texas	Significant negative bias; under predicted rutting.
Fatigue Cracking	Kansas	Sections 20-0101 and 20-0102 have a significant positive bias.	Colorado	All sections have a significant positive bias.
Fatigue Cracking	Iowa	Sections 19-0101 and 19-0102 have a significant positive bias.	Missouri	No fatigue cracking measured and none predicted.
Fatigue Cracking	Oklahoma	No fatigue cracking measured and none predicted.	Oklahoma	Section 0508 has a significant positive bias; all other sections—no fatigue cracking measured and none predicted.
Fatigue Cracking	Nebraska	Sections 31-0113 and 31-0114 have a significant positive bias.	Texas	No fatigue cracking measured and none predicted.
Thermal or Transverse Cracking	Kansas	Negative bias; under predicted transverse cracking.	Colorado	Significant negative bias; under predicted transverse cracking.
Thermal or Transverse Cracking	Iowa	Significant negative bias; under predicted transverse cracking.	Missouri	Small negative bias; minor lengths of transverse cracking measured and none predicted.
Thermal or Transverse Cracking	Oklahoma	Negative bias; under predicted transverse cracking.	Oklahoma	Negative bias; under predicted transverse cracking.
Thermal or Transverse Cracking	Nebraska	No transverse cracking measured and none predicted.	Texas	Significant negative bias; under predicted transverse cracking.

Rut Depth

- Rut depths were under predicted (negative bias) for the SPS-5 projects (refer to Figure A2-17), while no significant model bias exists for the SPS-1 projects.
- Rutting is the unbound layers and subgrade as high as 0.4 in. was calculated for some of the sections with measured rut depths less than 0.4 in. Thus, rutting in the unbound layers is probably being over predicted by the MEPDG. The amount of rutting in the unbound layers can be adjusted through the local calibration factors for fine and coarse grained soil—the β_{s1} term in the subgrade rutting screen. Independent adjustments for crushed stone aggregate layers and the subgrade can not be made. Thus, the local calibration value was determined for one layer. That layer was selected based on information extracted from the construction records for the specific project.
- It has been reported that the β_{s1} coefficient is related to the in place water content and density of the unbound layer at construction. Those sections with the highest measured rut depths were those with the higher water contents and lower densities in comparison to the maximum dry unit weight and optimum water content. Construction delays were also reported for some of the sections/projects (refer to Attachment A2.4.C).

Fatigue Cracking

- Fatigue cracking was under predicted for all test sections that exhibited fatigue cracking (refer to Figure A2-18). The test sections that exhibited the higher areas of fatigue cracking were found to have high and low air voids, as well as variable asphalt contents, in the HMA mixtures. No definite correlation or correspondence was identified between the residual errors and volumetric properties, design features, or site conditions.
- The Colorado SPS-5 project was the only HMA overlay to exhibit appreciable fatigue cracking. The MEPDG did not predict the level of cracking exhibited using the global calibration values and assuming full friction or bond between the HMA overlay and existing surface. However, the Colorado project was the only SPS-5 project used within this demonstration that included the placement of a thin leveling course prior to overlay placement. This leveling course may or may not be related to the occurrence of fatigue cracking.

Thermal Cracking

- The amount of thermal cracking is under predicted (refer to Figure A2-19). The default indirect tensile strength and creep compliance values for the wearing surface may or may not adequately account for the actual aging and varying volumetric properties. The test sections that exhibited the higher transverse cracking lengths were those with the higher air voids and/or lower asphalt contents.
- More importantly, thermal cracking is heavily dependent on the aging of the asphalt during production and construction. Some of the penetration and viscosity values measured on the recovered asphalt indicate brittle asphalt in comparison to the asphalt grade specified. In fact, two of the SPS-5 projects exhibited extensive transverse cracking in a warmer climate than Kansas (Texas and Oklahoma SPS-5 projects). The MEPDG did not predict any thermal cracking for these projects. The construction records indicate plant problems, mixture design issues, and severe hardening of the asphalt during the production process—high asphalt viscosities.

- The length of thermal cracks is adjusted through the local calibration factor for the wearing surface—the B_{t3}, B_{t2}, or B_{t1} terms in the thermal fracture screen. For this demonstration, only Input Level 3 was used for the material properties for predicting the length of thermal cracks.

IRI or Roughness

- The IRI values are heavily dependent on the other distresses calculated by the MEPDG and the site factor. Changing the local calibration factor from unity will affect the IRI values. Thus, the IRI predictions should be evaluated for bias only after the bias has been removed from the other prediction models.

Step 8—Eliminate Local Bias of Prediction Models

The process used to eliminate or reduce model bias depends on the cause of that bias and the accuracy desired by the agency. There are three possibilities, which were included under Step 8 for Demonstration 1. The three possibilities are reiterated below for ease of reference and consistency.

1. The residual errors are, for the most part, always positive or negative with a low standard error of the estimate in comparison to the trigger value, and the slope of the residual errors versus predicted values is relatively constant and close to zero. The precision of the prediction model is reasonable but the accuracy is poor. In this case, the local calibration coefficient is used to reduce the bias. This condition generally requires the least level of effort and the fewest number of runs or iterations of the MEPDG to reduce the bias.

2. The bias is low and relatively constant with time or number of loading cycles, but the residual errors have a wide dispersion varying from positive to negative values. The accuracy of the prediction model is reasonable, but the precision is poor. In this case, the coefficient of the prediction equation is used to reduce the bias but the value of the local calibration coefficient is probably dependent on some site feature, material property, and/or design feature included in the sampling template. This condition generally requires more runs and a higher level of effort to reduce dispersion of the residual errors.

3. The residual errors versus the predicted values exhibit a significant and variable slope that appears to be dependent on the predicted value. The precision of the prediction model is poor and the accuracy is time or number of loading cycles dependent—there is poor correlation between the predicted and measured values. This condition is the most difficult to evaluate because the exponent of the number of loading cycles needs to be considered. This condition also requires the highest level of effort and many more runs to reduce bias and dispersion.

The second and third points apply to this demonstration. An ANOVA was completed to determine whether e_r, b_o, and/or m are dependent on some factor included in the sampling matrix for some other design feature and site condition factor. No definite correlation was found between the statistical parameters and other parameters, but correspondence was observed between the residual error and selected layer properties for specific transfer functions.

To try and improve on the accuracy of the transfer function, the agency needs to first decide on whether to use the agency specific values or the local calibration factors that are considered as inputs in the MEPDG software. Either one can be used with success—for this demonstration the local calibration parameters were used. Time-history plots of each performance indicator should be prepared to determine if one or multiple local calibration parameters need to be evaluated, as noted above (refer to Attachment A2.4.D).

Table A2-14 summarized the results from using the global calibration values for predicting the performance indicators. An evaluation of the predictions and residual errors within and between the LTPP SPS projects was completed to reduce the bias. The process to estimate the agency or local calibration values is based on a review of the residual errors, and slope and intercept estimators in accordance with the sampling matrix—within and between the LTPP test sections. Guidance for reducing the bias and standard error is based on the following observations, analyses, and comparisons of the data.

Rut Depth Transfer Function

Even though the bias is small for new flexible pavement construction (0.0059), there is significant dispersion between the predicted and measured total rut depths—poor correlation between the measured and predicted values (refer to Figure A2-17). More importantly, the residuals for new construction are different than those for the HMA overlays. Thus, the local calibration parameters were used to reduce the residual error for the predicted rut depths. The bias of the rut depth prediction model can be adjusted through three parameters; two for the HMA layers (β_{r1}, and β_{r3}) and one for the unbound layers and subgrade (β_{s1}).

- **Unbound Layers and Soils—Local Calibration Parameter:** As noted above, it is highly probable that the rutting predicted in the unbound layers for the thick, deep-strength HMA pavements is being over predicted because the measured rut depths are less than the value predicted for the unbound layers and subgrade. Conversely, it is expected that the rut depths are under predicted for the thinner conventional flexible pavements with an aggregate base and/or subgrade that became wet as a result of rain delays during construction. Those sections with the highest measured rut depths were those with the higher water contents and lower densities, as well as for projects where construction delays occurred.
  - The amount of rutting in the unbound layers is adjusted through the local calibration parameters for fine and coarse grained soil—the β_{s1} term in the subgrade rutting screen. It has been reported that the β_{s1} coefficient is related to the in-place water content and density of the unbound layer at construction. Independent adjustments for aggregate base layers and coarse-graded subgrade soils, however, can not be made. The local calibration value was determined for one unbound layer. That layer was selected based on information extracted from the construction records for the specific project. An analysis of the rut depth residual errors resulted in the following observations, actions taken, and local calibration values that were used to reduce the bias for the LTPP SPS projects.
  - A rut depth of 0.1 in. was assumed for the unbound layers and soils constructed in accordance with the specifications under thick or deep-strength SPS sections in characterizing the existing condition of the pavement structure. A rut depth of 0.1 in. is considered to be immeasurable.This assumption was used to estimate the local calibration value for the unbound materials and layers of the thick and deep-strength flexible pavements with the lower measured rut depths.
  - The sections with higher water contents and lower densities in the unbound layers and/or sites that exhibited construction difficulties because of wet weather during construction exhibited much higher measured rut depths than predicted with the MEPDG.
  - For the sections with optimum or below optimum water contents of the unbound layers, measured rutting was consistently over predicted but at a much reduced level in comparison to those with higher water contents. The β_{s1} local calibration parameter of the unbound

layers was estimated based on water content and density in relation to the maximum dry density and optimum water content of the unbound materials and soils, which are provided below in Table A2-15.

Table A2-15. Local Calibration Parameter of Unbound Layers

Condition of Unbound Layers	Unbound Base	Subgrade Soil
Near maximum unit weight and optimum or equilibrium water content	0.5	0.5
Near maximum unit weight but higher water contents above optimum	0.75	0.75
Lower densities (below maximum unit weight) and higher water contents (above optimum)	3.0	1.25

Insufficient data were available to determine if the rut depth local calibration parameter values were dependent on the type of soils; coarse versus fine-grained soils.

- **HMA Layers and Mixtures**—An analysis of the residual errors resulting from use of the global calibration values (refer to Figure A2-17 and Table A2-14) found that the HMA local calibration parameter was related to air voids and asphalt content as shown in Table A2-16. An analysis of the residual errors resulted in the following observations, actions taken, and local calibration values that were used to reduce the bias for the LTPP SPS projects.

Table A2-16. An Analysis of Residual Errors from the Use of Global Calibration Values

Asphalt Content (Typical Range)	Local Calibration Parameters		Air Voids (Typical Range)		
			High	Medium	Low
High	β_{r1}	Coefficient	1.5	1.5	1.25
	β_{r3}	Exponent	1.05	1.15	1.15
Medium	β_{r1}	Coefficient	1.5	1.30	1.25
	β_{r3}	Exponent	0.95	1.0	1.05
Low	β_{r1}	Coefficient	1.6	1.35	1.35
	β_{r3}	Exponent	0.9	1.0	1.0

Note: Insufficient data to produce specific relationships between the volumetric properties and the local calibration parameters, so a general range was used for the properties.

  - The reported plant problems and mixture design issues mentioned in the construction reports resulted in confounding factors that had a significant effect on the local calibration values. These confounding factors complicate the local calibration process and make it difficult to increase the precision and accuracy of the rut depth prediction model. The

LTPP SPS projects with confounding factors were analyzed and treated separately (SPS-5 projects—Oklahoma and Texas; SPS-1 project—Iowa).

- The slope of the rut depth with time relationship was project specific but found to be related to the volumetric properties of the mixture.
- Rutting was consistently under predicted for most of the LTPP SPS-5 HMA overlays. No significant difference in rutting was found between the HMA mixtures with and without RAP for the LTPP SPS-5 projects.

The local calibration values were used to predict the rutting of all test sections included within this demonstration. Table A2-17 provides a summary of the statistical parameters resulting from use of the local calibration values for HMA and the unbound layers. Figure A2-21 provides a comparison of the predicted and measured rut depths for the new construction (SPS-1 projects) and HMA overlays (SPS-5 projects). As shown, there is an increase in the precision of the transfer function, especially for new construction projects.

Table A2-17. Summary of the Statistical Parameters—Local Calibration Values Used for Predicting

Performance Indicator	Project	Bias	Standard Error	s_e/s_y	R^2	Hypothesis; H_o: $y_i - x_i = 0$	Comment
Rutting	SPS-1	+0.028	0.134	0.569	0.676	Accept	The rut depths measured on the SPS-5 projects were significantly lower than those measured on the SPS-1 projects.
	SPS-5	−0.031	0.0642	0.758	0.357	Accept	
Fatigue Cracking	SPS-1	+1.363	14.333	0.851	0.360	Accept	Correlation is still considered poor because of construction anomalies.
	SPS-5	−1.150	4.900	0.695	0.683	Accept	
Transverse Cracking	SPS-1	+23.53	287.4	0.603	0.696	Accept	It is expected that the reported plant problems and mixture design issues result in a positive or negative bias, depending on the value used.
	SPS-5	−100.1	606.6	0.672	0.639	Accept	
IRI	SPS-1	+2.804	14.348	0.571	0.668	Accept	Correlation is still considered poor because of construction anomalies.
	SPS-5	+4.900	14.077	0.959	0.121	Accept	

Fatigue Cracking Transfer Function

Measured fatigue cracking was under predicted for almost all test sections that exhibited fatigue cracking (both new construction and HMA overlays; refer to Figure A2-15). However, only three of the eight LTPP SPS projects exhibited high levels of fatigue cracking, and construction problems/delays occurred for all three projects (SPS-1 projects—Iowa and Kansas; and SPS-5 project—Colorado). Specific test sections within some of the other LTPP SPS projects also exhibited higher levels of fatigue cracking. That negative bias from all test sections should be reduced, if possible.

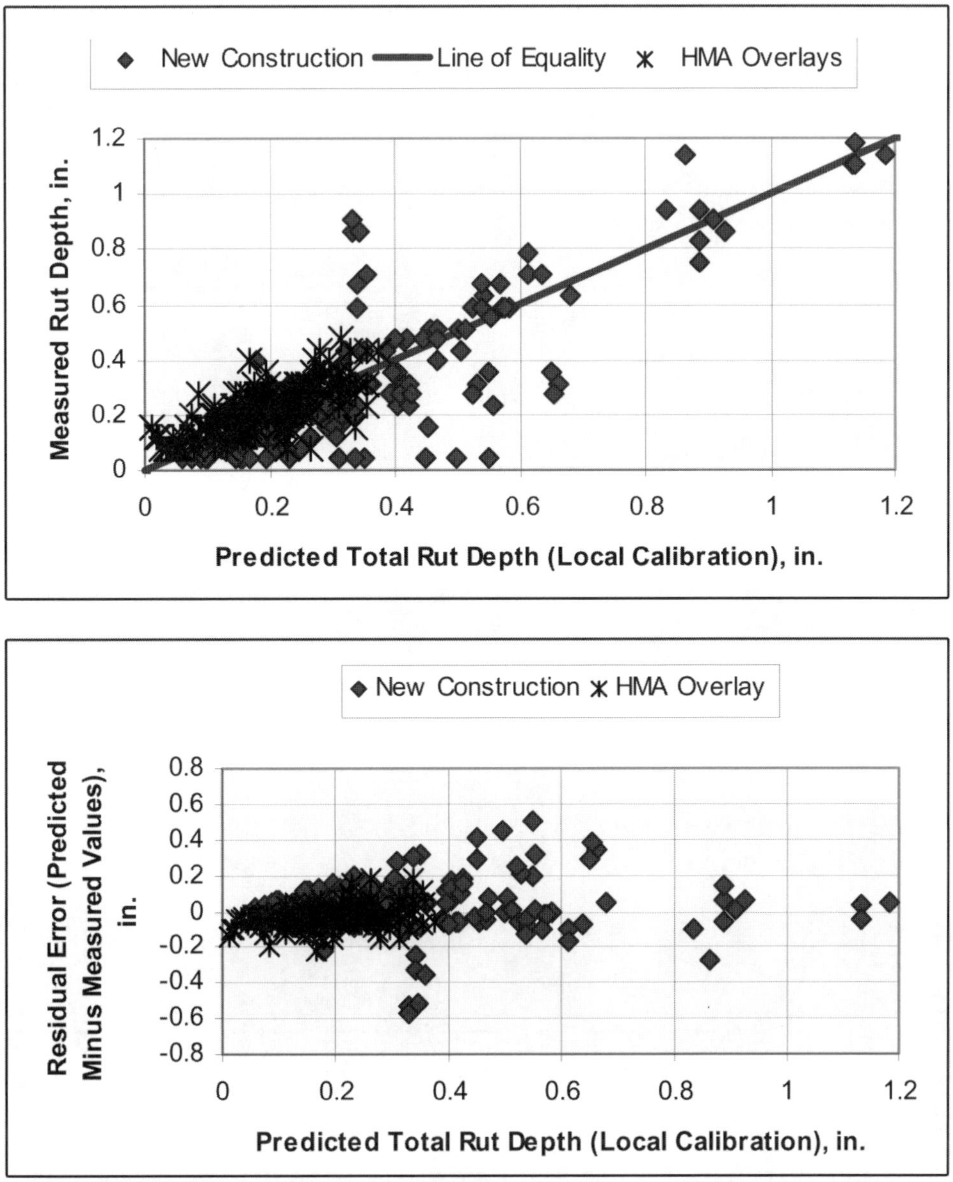

Figure A2-21. Comparison of Predicted and Measured Rut Depths Using the Local Calibration Values for the Subgrade, Unbound Aggregate, and HMA Layers

The bias of the fatigue cracking prediction model can be adjusted through four parameters; β_{f1}, β_{f2}, β_{f3}, and C_2. The β-terms are related to calculating the allowable number of load applications for a specific condition and layer, while the C_2-term is related to calculating the percent area of fatigue cracking from the damage index. The β_{f1}- and C_2-terms are the ones typically used to eliminate the model bias and/or reduce the standard error of the estimate. The other two local calibration parameters are assumed to be adequate.

The C_2 parameter in the bottom-up fatigue cracking prediction equation was also excluded from the analysis. It was assumed that the C_2 value of unity determined from the global calibration process was appropriate for the LTPP SPS projects considered within this demonstration. That assumption for C_2, however, is probably incorrect, just like for the Kansas PMS segments. The growth in fatigue cracking with time can be much steeper than predicted by the MEPDG using the global calibration value of unity. This condition is illustrated in Figure A2-22 for some of the LTPP SPS-1 and SPS-5 test sections with the higher amounts of fatigue cracking. This difference between the measured and predicted values with time decreases the precision of the fatigue cracking prediction model for the LTPP projects. However, there are too few LTPP SPS projects and individual test sections without anomalies and with appreciable amounts of fatigue cracking to determine a reliable estimate of C_2, similar to the finding for the Kansas PMS segments.

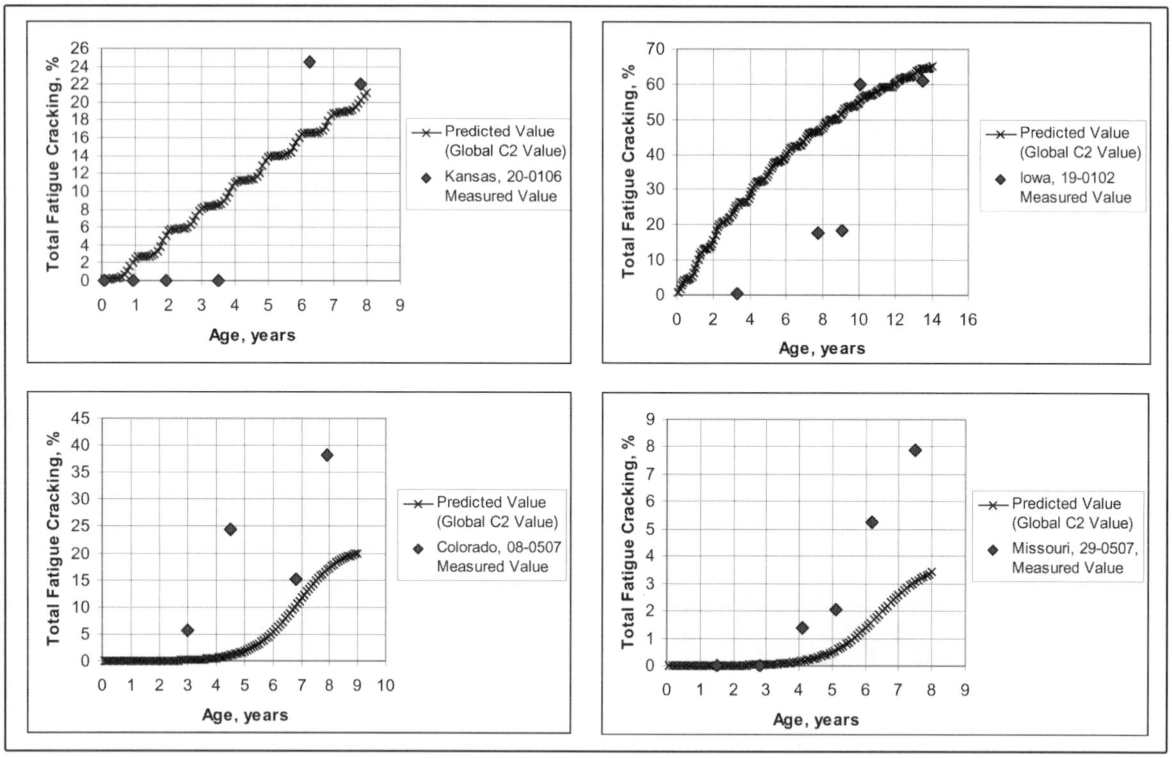

Figure A2-22. Comparison of Measured and Predicted Values of Fatigue Cracking for Specific Test Sections

An analysis of the residual errors for the test sections that exhibited the higher areas of fatigue cracking was completed to identify the material properties or site features related to the residual error. No clear correlation was identified, probably because of the confounding factors and higher number of anomalies between the LTPP SPS projects (refer to Attachment A2.4.C). Thus, a constant value was used to predict the fatigue cracking for the local calibration β_{f1}-term, which is 0.005.

In order for the residual error to be minimized for the SPS-5 Colorado project, the local calibration term would have to be much lower than the above value. This LTPP SPS-5 project was the only one with a thin leveling course. Assuming zero interface friction or bond between the existing surface and HMA overlay results in a local calibration value of 0.005 for this project. No significant difference in cracking was found between the HMA mixture with and without RAP.

Table A2-17 provides a summary of the statistical parameters resulting from use of the fatigue cracking local calibration value. Figure A2-23 provides a comparison of the predicted and measured fatigue cracking for new construction (SPS-1 projects) and HMA overlays (SPS-5 projects). As shown, there is an increase in the accuracy of the transfer function (Figure A2-18 compared to Figure A2-23), but the correlation between the predicted and measured values is still considered poor for the LTPP SPS-1 projects (i.e., the MEPDG is considered accurate but has poor precision based on the LTPP projects included for this demonstration).

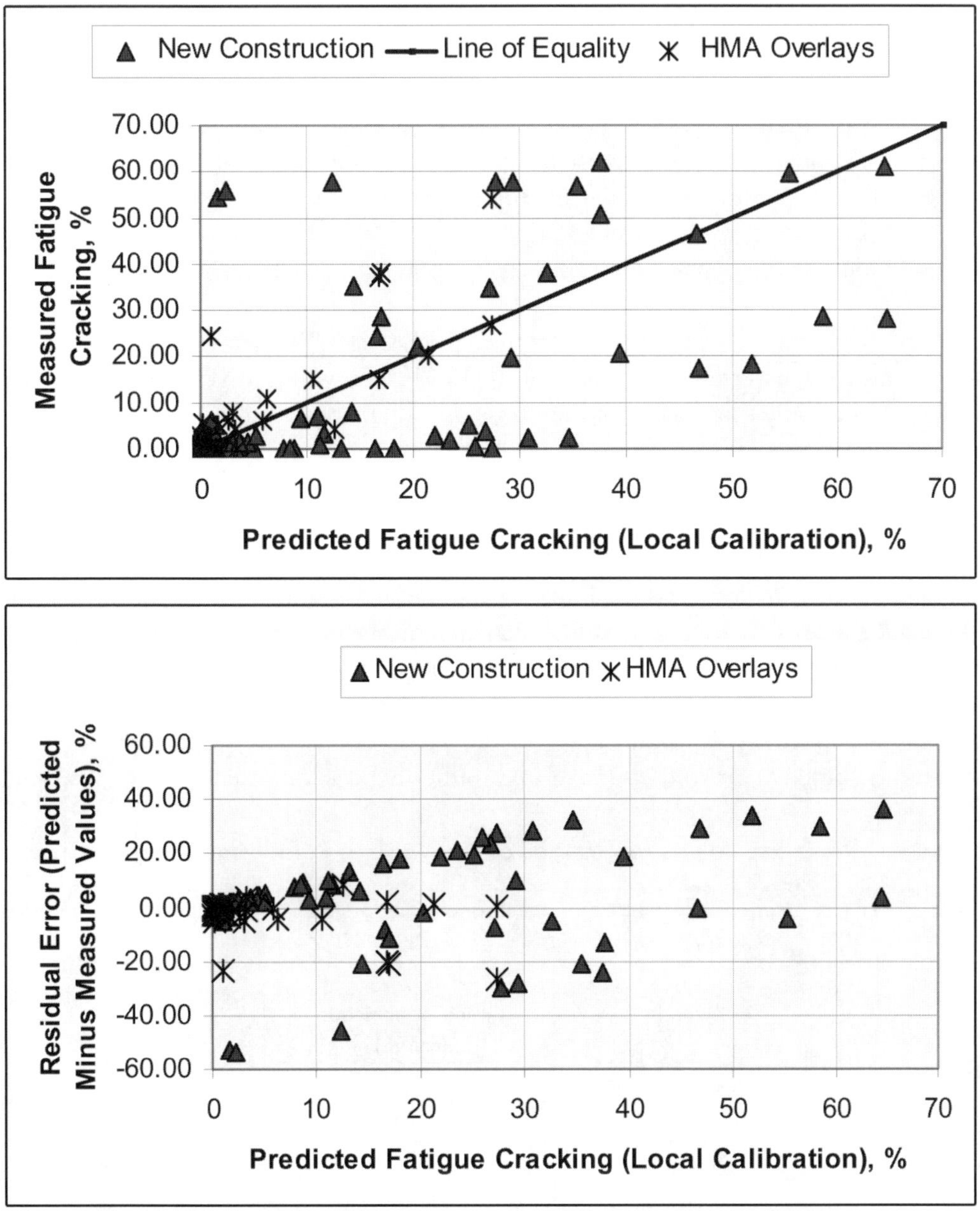

Figure A2-23. Comparison of Predicted and Measured Fatigue Cracking Using a Local Calibration Value for the HMA Mixture That Is Air Void Dependent

Reasons for this poor correlation are believed to be the result of different construction problems or anomalies that occurred on all of the LTPP SPS-1 projects selected for this demonstration. Variable support conditions of the unbound layers and problems that occurred during HMA production (resulting in hard to brittle HMA mixtures) would have a significant effect on the fatigue resistance of the pavement structure. On a positive note, however, the HMA-mixture production problems that severely hardened the HMA mixtures should result in low local calibration values in comparison to the global calibration values of unity, which they do.

Although the hypothesis for the condition of using the local calibration values was accepted, these values should not be used from a practical engineering standpoint because they are heavily influenced by severely hardened or aged HMA.

Transverse Cracking Transfer Function

The length of transverse or thermal cracks was significantly under predicted for nearly all of the test sections for new construction and HMA overlays using the global calibration values (refer to Figure A2-19). Only one of the SPS-1 sites had significant lengths of transverse cracking—the Iowa project. Most of the SPS-5 projects, however, did exhibit various levels of transverse cracking. Thus, the MEPDG prediction model resulted in a significant negative bias, and that bias should be eliminated.

The maximum length of thermal cracks predicted by the MEPDG is 2,200 ft/mi, which corresponds to about a 30-ft spacing of transverse cracks. Many of the LTPP sites with transverse cracks exceed that maximum limit. Thus, only those measured responses less than about 2,500 ft/mi should be used in the local calibration process.

Thermal cracking was under predicted for nearly all of the test sections, and the magnitude of the residual error did correspond to the magnitude of the air voids of the wearing surface. The HMA-mixture production and mixture design issues, however, resulted in significant confounding factors that reduced the significance of air voids and asphalt contents on the occurrence of transverse cracking. These construction problems are considered anomalies within this demonstration. The β_{t3} local calibration parameter was estimated based on the air voids and production problems of the wearing surface and used to reduce the model bias (refer to Table A2-13). The local calibration values (β_{t3}) determined for the different air void ranges related to thermal cracking are listed below.

Table A2-18. Local Calibration Values for Ranges of Air Voids in Relation to Thermal Cracking

Plant Hardening of Asphalt	Range of Asphalt Contents, %	Range of Air Voids, %	Local Calibration Value, β_{t3}
Typical, no excessive hardening during production	>10 (high)	5 to 7 (typical specification range)	1.0
	8 to 10 (typical to higher asphalt contents)	7 to 9	1.7
		>10	2.0
	<8 (low)	>10	5.0
Excessive	Not important	Not important	7.5
Severe	Not important	Not important	20.0

Table A2-17 lists the thermal cracking bias using the local calibration values listed above. As shown, the hypothesis is now accepted for the new construction projects (SPS-1) but is still rejected for the HMA overlays (SPS-5). Figure A2-24 compares the predicted and measured thermal cracking using the local calibration values and shows an increase in the accuracy of the transfer function, as compared to use of the global calibration values (refer to Figures A2-19 and A2-24).

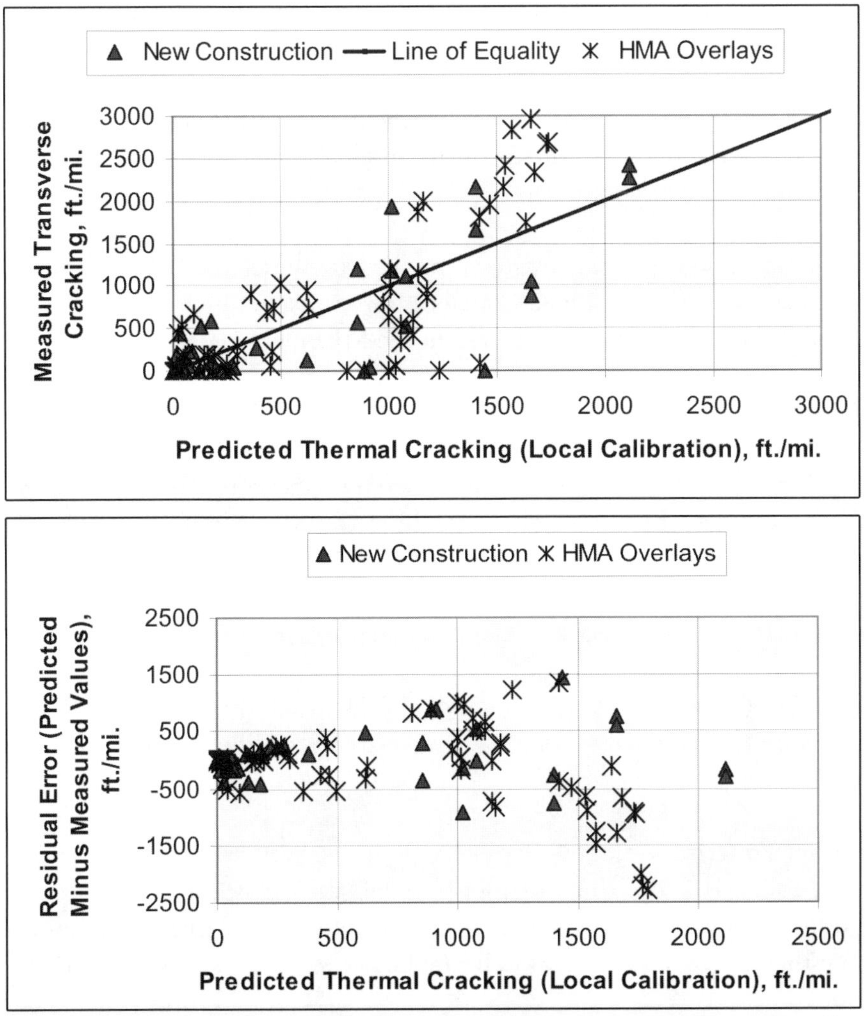

Figure A2-24. Comparison of Predicted Thermal Cracking and Measured Transverse Cracking Using the Local Calibration Values for the HMA Mixture

Roughness or IRI Regression Model

The IRI values predicted by the MEPDG using the global calibration values are within acceptable limits of the measured values. The hypothesis was accepted in that the bias is considered minimal (refer to Table A2-13 and Figure A2-20). Those IRI values, however, are heavily dependent on the other distresses predicted by the MEPDG. Any changes to the predicted distresses from the global calibration process will affect the IRI values. As an example, the test sections for which the IRI values were under predicted were for those sections where the transverse cracks were significantly under predicted. Greater predicted lengths of the thermal cracks using the local calibration values will result in higher IRI values being predicted by the MEPDG. Thus, the IRI model needs to be re-evaluated after the bias has been removed from the other prediction models.

IRI values were predicted with the MEPDG after all of the local calibration adjustments were made to the other distress prediction models. Table A2-17 summarizes the statistical values for the SPS-1 (new construction) and SPS-5 (HMA overlays) projects, while Figure A2-25 compares the predicted and measured IRI values. As shown, the hypothesis is accepted and the correlations are considered reasonable without bias. Some observations from this comparison and evaluation of the data are noted below.

- The MEPDG IRI prediction equation begins to over predict the measured IRI values for some of the test sections for longer times or older ages. These positive residual errors are probably caused by over predicting the other distresses discussed above.

- The MEPDG IRI prediction equation significantly under predicts the measured IRI for some of the new construction test sections. Other non-load related distresses that are not predicted by the MEPDG can affect the IRI values that are not considered in the MEPDG.

Step 9—Assess Standard Error of the Estimate

After the bias was reduced for each of the transfer functions, the SEE is evaluated over the range of predicted distress values. The SEE for the local calibration process and predictions is summarized in Table A2-17. Figure A2-26 compares the SEE for the globally calibrated transfer functions to the SEE for the locally calibrated transfer functions. Using the local calibration values, the SEE values were found to be similar to or greater than the SEE values included in the MEPDG software. The following summarizes the comparison of the values between the global and local calibration values.

- **Rut Depth Transfer Function (Total Rut Depth)**—Standard errors are lower from the local calibration in comparison to the global calibration, similar to the findings from Demonstration 1.

- **Alligator Cracking Transfer Function**—Standard error based on the local calibration is similar to the global values at the lower predicted values of fatigue cracking, but continues to increase to values significantly greater than the global values. A limit was placed on the SEE value from the global calibration, but that limit was found not to be applicable to the LTPP sites included within this demonstration.

- **Thermal Cracking Transfer Function**—Standard error based on the local calibration is consistently higher than the values determined from the global calibration process, similar to the findings from Demonstration 1.

- **IRI Regression Model**—Standard error for the IRI regression equations is not provided in the MEPDG software screens and cannot be changed.

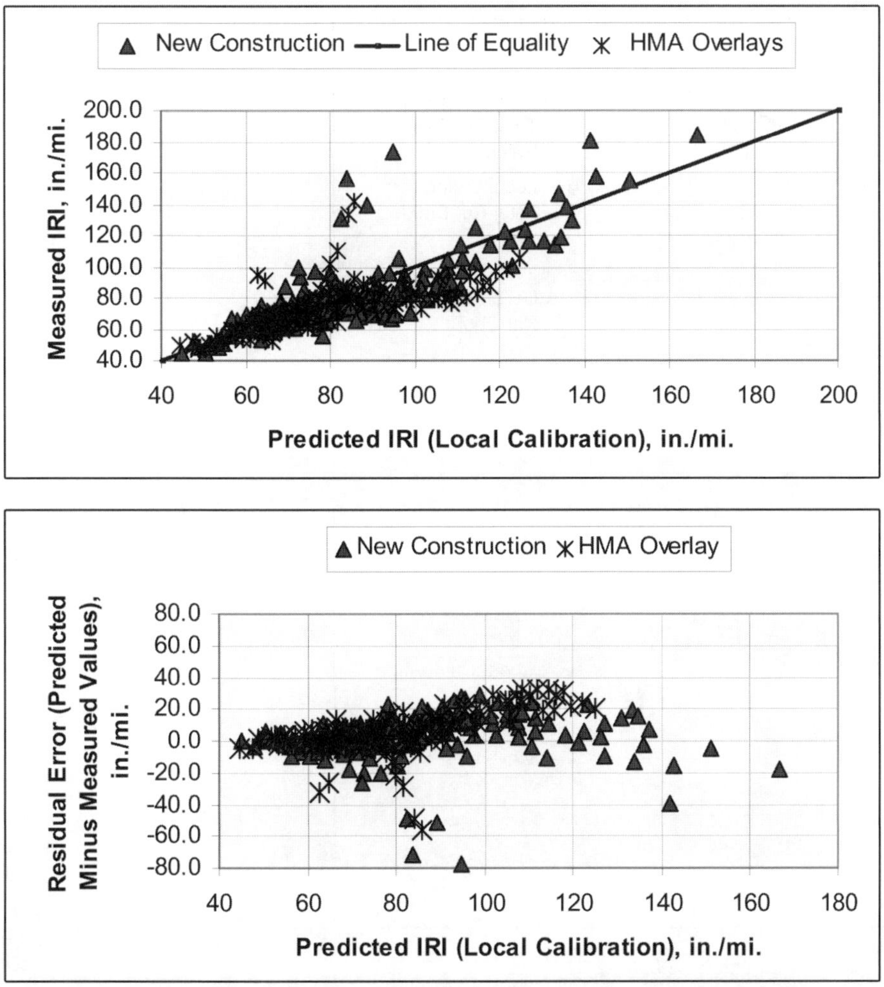

Figure A2-25. Comparison of Predicted and Measured IRI Values Using the Global Calibration Values

Step 10—Reduce Standard Error of the Estimate

As noted in Step 9 and shown in Figure A2-26, the SEE from the local calibration process was found to be different than the SEE relationships included in the MEPDG software for rutting, fatigue cracking, and thermal cracking. An ANOVA can be completed to determine if the residual error or bias is dependent on some other parameter or material/layer property for the LTPP test sections. No correlation was identified, so the SEE values shown in Figure A2-26 and the local calibration factors summarized in Step 8 are believed to be the final values for the LTPP test sections included in the sampling template. A possible reason that the values were not correlated is a result of the anomalies found at these LTPP sites. Thus, no further reduction in the SEE is possible based on the more simplistic evaluation.

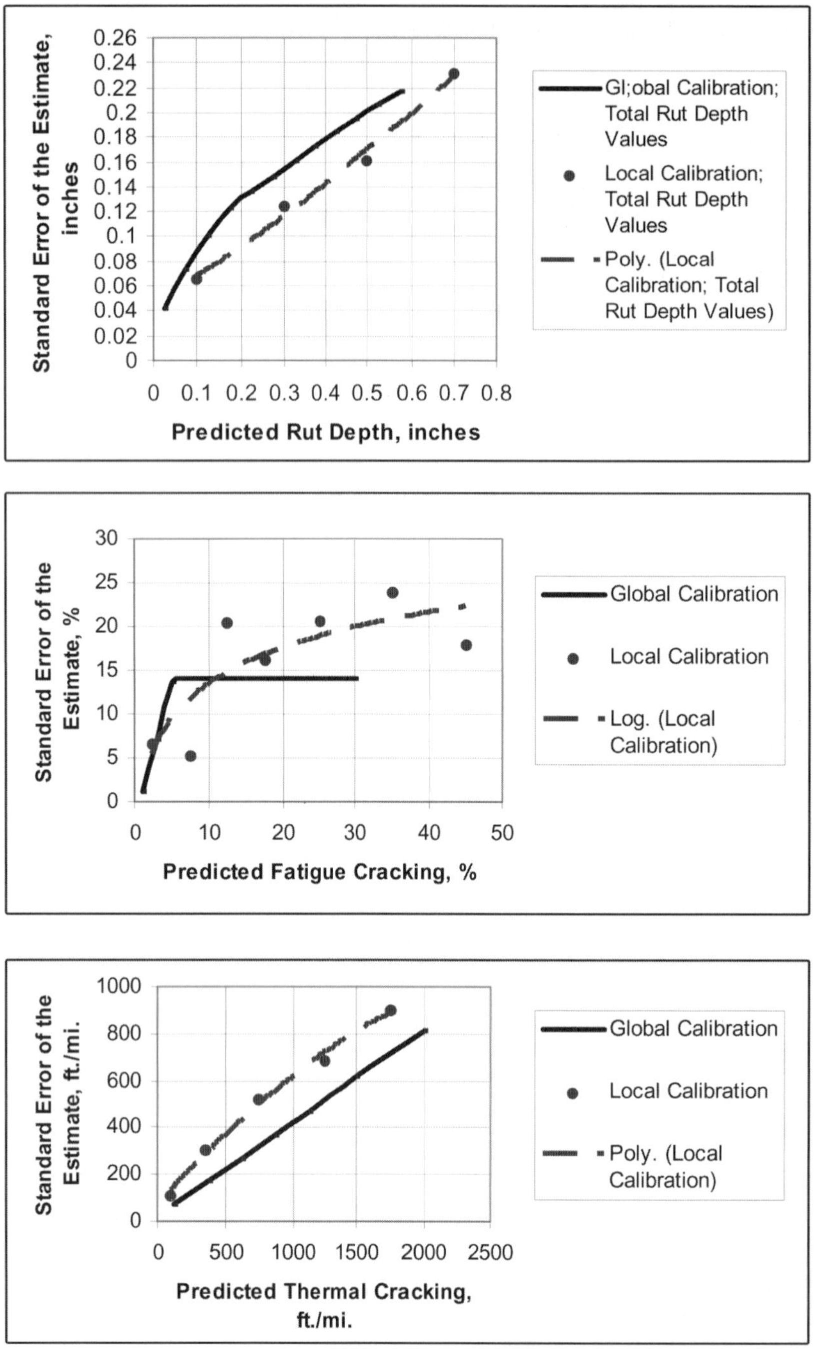

Figure A2-26. Comparison of the Standard Error of the Estimate from the Global and Local Calibration Process

Step 11—Interpretation of Results and Deciding on Adequacy of Calibration Factors

For this demonstration, the global calibration values did result in a bias for all distresses, with the exception of the IRI regression model. The MEPDG did not accurately explain the differences in performance between the different HMA mixtures and pavement structures. To reduce that bias required local calibration values that were significantly different from unity. The MEPDG IRI regression equation was the only model that was confirmed using data from selected LTPP SPS projects in Kansas and adjacent states. The purpose of this step is to decide whether to adopt the local calibration values or continue to use the global values that were based on data included in the LTPP program from around the United States.

As stated under Demonstration 1 using the Kansas PMS data, to make that decision an agency should identify major differences between the LTPP projects and the standard practice of the agency to specify, construct, and maintain its roadway network. The agency should also determine whether the local calibration values can explain those differences, and evaluate any change from unity for the local calibration parameters to ensure that the change provides engineering reasonableness.

The interpretation of results is discussed further in Section A2.3 (Summary for Local/Regional Calibration Values) using the two different data sets: PMS segments and selected LTPP SPS projects in and adjacent to Kansas. The following briefly interprets some of the results using the LTPP SPS data.

- The IRI regression equation was found to be a reasonable simulation of the IRI values measured on the LTPP test sections. This finding was expected because other LTPP test sections were used to develop the regression model. The IRI prediction equation is believed to be adequate for Kansas' climate, materials, and other site features.

- All HMA mixtures included in the LTPP SPS projects included in this demonstration are more susceptible to fracture than included in the global calibration process. These mixtures are brittle in comparison to those used to determine the global calibration values. Most of these sections that have exhibited higher amounts fatigue and transverse cracking are those where plant problems occurred that severely hardened the asphalt. The local calibration values determined from those projects would not be recommended for use for typical pavement design projects.

- The C_2 parameter is significantly different from unity (refer to Figure A2-21), but there were too few LTPP SPS projects with higher levels of fatigue cracking and without anomalies to determine a reliable estimate for this parameter. Thus, the global calibration value for C_2 (unity) should continue to be used until more projects are included in the local calibration process without construction anomalies and with higher amounts of fatigue cracking to confirm or dispute that observation.

- All mixtures are also more susceptible to thermal cracking than those included in the global calibration process. Substantial lengths of transverse cracking were exhibited on many of the LTPP SPS projects. Most of the projects with excessive transverse cracking were those that exhibited construction and mixture production problems. Other LTPP SPS projects were included without any known construction problems or anomalies. The local calibration value was significantly greater than unity. Thus, it would be recommended that the local calibration value for thermal cracking be used for design.

- The HMA mixtures with and without RAP in the LTPP SPS-5 projects did not exhibit any difference between the local calibration values for each distress. Thus, the local calibration values would be the same regardless of the percentages of RAP included in those mixtures. The percentage of RAP used in the HMA mixtures for the LTPP SPS-5 projects was generally less than 25 percent.

- The subgrade rutting local calibration value is believed to be reasonable because of the findings from previous forensic studies and would be recommended for use.
- The rut depth local calibration values for the HMA mixtures do deviate from unity. In summary, the HMA local calibration values for rutting would be recommended for use. The SEE values derived from the local calibration were also lower than the SEE values derived from the global calibration.

A2.3 Summary for Local/Regional Calibration Values

The summary of results is discussed in two parts, which is a further interpretation of the results from Step 11 of both demonstrations. The first part is to compare the results from the two demonstrations—the Kansas PMS segments and the LTPP test sections. The second part of the discussion relates to whether the results represent standard practice and should be used for designing flexible pavements and HMA overlays that are commonly used in Kansas.

A2.3.1 Comparison of Results: PMS Segments and LTPP SPS Test Sections

There are differences between the two demonstrations of the local validation-calibration process. Table A2-19 summarizes the more important differences, as well as similarities, that can have an effect on the outcome. Although substantial differences exists between the two data sets, the MEPDG transfer functions were found to be reasonably accurate after local calibration for all performance indicators using both the Kansas LTPP and adjacent state projects and the Kansas PMS segments. The PMS segments do exhibit higher within project variability, while the between project variability is greater for the LTPP test sections. As previously noted, this was expected because of the experimental design for the LTPP program.

The statistical parameters resulting from the two demonstrations were summarized in Tables A2-8 and A2-17. The precision of the transfer functions is less than desired based on the target values included in Table A2-5. The s_e/s_y term is relatively high and the R^2 term relatively low for most of the transfer functions. The reason for the lower precision is not necessarily the result of poor prediction models in the MEPDG or high lack-of-fit modeling errors. Making the transfer function more precise (reducing the s_e/s_y and increasing R^2) is not likely, as previously stated, because of the large measurement error within both data sets, especially for the PMS data set (refer to Attachment A2.4.B). Until the measurement precision of the performance indicators can be improved (reduction of the measurement error), it would be recommended that the SEE relationships included in the MEPDG for each transfer function continued to be used for Kansas, with the exception of the rut depth function. The SEE values for the rut depth transfer function were found to be consistently lower than the values derived from the global calibration process. The following summarizes and compares the local calibration values for the MEPDG distress transfer functions that were determined from the two demonstrations.

Table A2-19. Comparison of the Two Demonstrations for Flexible Pavements and HMA Overlays

LTPP SPS-1 and SPS-5 Sections	Kansas PMS Segments
Sampling Template Balanced with replication. 8 projects, 56 sections.	**Sampling Template** Unbalanced without replication. 16 projects, 16 segments.
Data Collection Frequency/Observations 0.3 to 2 per year. 196 observations	**Data Collection Frequency/Observations** One per year. 112 observations.
Distress Data • No adjustments needed for the distresses to be consistent with the MEPDG. • Large measurement errors present in the distress data.	**Distress Data** • Distresses must be adjusted to be consistent with MEPDG predictions. • Large measurement errors present in the distress data.
Truck Traffic Moderate levels of AADTT.	**Truck Traffic** Low to moderate levels of AADTT.
Climate Projects located in Kansas and adjacent states; both warmer and cooler areas.	**Climate** Projects restricted to locations in Kansas.
Pavement Structures • Conventional, deep-strength, and full-depth sections; sections include granular layers and ATB. • HMA overlays of flexible pavements.	**Pavement Structures** • Full-depth HMA pavement; no granular layers and no ATB. • HMA overlays of flexible pavements.
HMA Mixtures Dense-graded conventional neat HMA mixtures with and without RAP included in the sampling template.	**HMA Mixtures** Dense-graded conventional neat HMA, PMA, and Superpave mixtures included in the sampling template.
The individual sections represent different mixture and structural characteristics for the same traffic level—different mixture and structural design procedures and features used.	Individual sections represent the same mixture and structural characteristics for the segment traffic—same mixture and structural design procedures were used.
Construction Anomalies Many of the projects and test sections had various anomalies that have a significant impact on performance.	**Construction Anomalies** None identified in database; all layers believed to be constructed in accordance with project specifications.
Data from some overlay sections indicate no bond between the overlay and existing surface.	Data from some overlay sections indicate no bond between the overlay and existing surface.

A2.3.1.1 Alligator (Fatigue) Cracking Transfer Function

Local calibration values were determined from both the Kansas PMS and LTPP data. Results from both data sets found that the HMA mixtures have inferior fracture properties or much more cracking, than those included in the global calibration database. In summary, the Kansas PMS and LTPP SPS data sets resulted in comparable values for the local calibration parameters for fatigue cracking. This similar finding provides some confidence that the transfer function included in the MEPDG is appropriate for the conditions and materials commonly used by KSDOT. A summary of the values are listed in Table A2-20.

Table A2-20. HMA Layers/Mixture Local Calibration Parameters, β_{f1}, C_2

<table>
<tr><th colspan="3">Kansas PMS Segments</th><th colspan="2">LTPP Sections;
Kansas and Adjacent States</th></tr>
<tr><td colspan="3">Constructed in accordance with project specifications assumed for all projects.</td><td colspan="2">Deviations from project specifications for the HMA layers on multiple projects.</td></tr>
<tr><td colspan="3">Data from some overlay sections indicate no bond between the HMA overlay and existing surface.</td><td colspan="2">Data from some overlay sections indicate no bond between the HMA overlay and existing surface.</td></tr>
<tr><td colspan="5">Local Calibration Value for the allowable number of load cycles coefficient; β_{f1}</td></tr>
<tr><td>HMA Overlays</td><td>All Mixes</td><td>0.05</td><td>HMA Overlays;
All Mixes</td><td>0.005</td></tr>
<tr><td rowspan="3">New Construction; Mixture Dependent</td><td>Conventional Dense-Graded Neat HMA Mix</td><td>0.05</td><td rowspan="3">New Construction; All Mixes</td><td rowspan="3">0.005</td></tr>
<tr><td>Superpave</td><td>0.005</td></tr>
<tr><td>PMA</td><td>0.0005</td></tr>
<tr><td colspan="5">Local Calibration Value for the HMA-fatigue cracking coefficient; C_2</td></tr>
<tr><td colspan="3">The C_2 parameter probably has a value around 5.0 based on a limited number of PMS segments with appreciable amounts of fatigue cracking.</td><td colspan="2">The C_2 parameter probably has a value around 4.0 based on a limited number of LTPP SPS sections with appreciable amounts of fatigue cracking and without construction anomalies.</td></tr>
</table>

It should be noted that one reason for the less fracture resistant mixtures could be the result of lack-of-bond between adjacent HMA layers and accelerated aging of the HMA mixtures caused by more production hardening than simulated in the mathematical aging model included in the MEPDG. If mixture properties were available for input Levels 1 and 2, the local calibration values may have been closer to unity.

A2.3.1.2 Rut Depth Transfer Function

Similar local calibration values for the rut depth prediction model were also obtained from both the Kansas PMS and LTPP data and provide some confidence that the rut depth transfer function is applicable to the conditions and mixtures commonly used in Kansas. A summary of the values are listed below.

Table A2-21. Subgrade/Unbound Layer Local Calibration Parameter, β_{s1}

Kansas PMS Segments		LTPP Sections; Kansas and Adjacent States		
Constructed in accordance with project specifications assumed for all projects.		Deviations from project specifications for the unbound layers on selected projects.		
HMA Overlays	0.5	HMA Overlays	0.5	
New Construction	0.5	New Construction, Condition Dependent:		
		Near max. unit weight and optimum water content	0.5	
		Near max. unit weight, but higher water content	0.75	
		Lower densities and higher water content	Unbound Base	Subgrade
			3.0	1.25

Table A2-22. HMA Layers/Mixture Local Calibration Parameters, β_{r1}, β_{r3}

Kansas PMS Segments			LTPP Sections; Kansas and Adjacent States			
Constructed in accordance with project specifications assumed for all projects.			Deviations from project specifications for the HMA layers on multiple projects.			
Data from some overlay sections indicate no bond between the HMA overlay and existing surface.			Data from some overlay sections indicate no bond between the HMA overlay and existing surface.			
Local Calibration Value for the HM-Rut Depth coefficient; β_{r1}						
HMA Overlays	All Mixes	1.5	HMA Overlays and New Construction, Mix Property Dependent			
New Construction; Mixture Dependent	Conventional Dense-Graded Neat HMA Mix	1.5	Asphalt Content, Typical Range	Air Void, Typical Range		
				High	Medium	Low
			High	1.5	1.5	1.25
	Superpave Mix	1.5	Medium	1.5	1.30	1.25
	PMA Mix	2.5	Low	1.6	1.35	1.35
Local Calibration Value for the HMA-Rut Depth Exponent to the Number of Load Applications; β_{r3}						
HMA Overlays	All Mixes	0.95	HMA Overlays and New Construction, Mix Property Dependent			
New Construction; Mixture Dependent	Conventional Dense-Graded Neat HMA Mix	0.90	Asphalt Content, Typical Range	Air Void, Typical Range		
				High	Medium	Low
			High	1.05	1.15	1.15
	Superpave Mix	1.20	Medium	0.95	1.0	1.05
	PMA Mix	1.15	Low	0.90	1.0	1.0

A2.3.1.3 Thermal Cracking Transfer Function

Similar local calibration values for the thermal cracking transfer function were obtained from both the Kansas PMS and LTPP data. A summary of the β_{t3} values are listed below in Table A2-23.

Table A2-23. Summary of β_{t3} Values from Kansas PMS and LTPP Data

<table>
<tr><th colspan="3">Kansas PMS Segments</th><th colspan="4">LTPP Sections; Kansas and Adjacent States</th></tr>
<tr><td colspan="3">Constructed in accordance with project specifications assumed for all projects.</td><td colspan="4">Deviations from project specifications for the HMA layers on multiple projects.</td></tr>
<tr><td>Mixtures</td><td>New Construction</td><td>Overlays</td><td>Plant Hardening</td><td>Asphalt Content Range</td><td>Air Void Range</td><td>Local Cal. Value</td></tr>
<tr><td rowspan="2">PMA</td><td rowspan="2">2.0</td><td rowspan="2">2.0</td><td rowspan="4">Typical from plant production</td><td>High, >10</td><td>Typical, 5–7</td><td>1.0</td></tr>
<tr><td rowspan="2">8 to 10</td><td>7 to 9</td><td>1.7</td></tr>
<tr><td rowspan="2">Conventional Dense-Graded Neat HMA</td><td rowspan="2">2.0</td><td rowspan="2">7.5</td><td>High, >10</td><td>2.0</td></tr>
<tr><td>Low, <8</td><td>High, >10</td><td>5.0</td></tr>
<tr><td rowspan="2">Superpave</td><td rowspan="2">3.5</td><td rowspan="2">7.5</td><td>Excessive</td><td rowspan="2" colspan="2">Not Important</td><td>7.5</td></tr>
<tr><td>Severe</td><td>20.0</td></tr>
</table>

A2.3.1.4 IRI Regression Model

The IRI global calibration values were confirmed using both the Kansas PMS segments and the LTPP SPS projects. Both data sets resulted in reasonable correlation between the predicted and measured IRI values, excluding the projects with construction anomalies. The construction anomalies identified for some of the LTPP SPS-5 projects resulted in poorer correlations (refer to Table A2-17). Thus, it would be recommended that the IRI global calibration values and SEE relationship included in the MEPDG be used for pavement design.

A2.3.2 Application of Results from Local Calibration Process for Pavement Design

In deciding whether to adopt or use the results from the local validation-calibration process; these three questions need to be answered:

1. Do the local calibration values significantly differ from unity? The two demonstrations using different data sets reach the same conclusion—the local calibration values do differ from unity for all performance indicator transfer functions, with the exception of the IRI regression model.

2. If the values are different from unity, which they are for these demonstrations, why are they different? Reasons for the difference are unknown based on the information currently available. However, the MEPDG was globally calibrated using predominately data from the LTPP database, mostly LTPP-GPS test sections. The LTPP-GPS test sections represent the better performing pavements across the United States. The local calibration values support that hypothesis.

3. Do the local calibration values represent standard practice of the agency? The individual agencies have to answer this question. The user agency should also use the global and local calibration values to determine the magnitude of changes in the predicted performance indicators for typical site conditions and design features (traffic, climate, materials, etc.). Based on the two sets of data, it would be assumed that the results from the Kansas PMS segments do represent the standard practice of the Kansas DOT. Obviously, the construction anomalies identified for some of the LTPP SPS projects previously summarized would not represent standard practice of KSDOT. It

would not be reasonable to use results from projects with known construction anomalies for day-to-day designs.

The values determined for the conditions where the layers were not constructed in accordance with the project specifications should not be used for design. These values, however, could be used for special forensic studies and investigations to identify probable cause of premature failures. Based on the findings from both demonstrations, the following local calibration values would be recommended for use in day-to-day designs within this region. It is believed that there is sufficient data to support their use, even though use of these values will result in more predicted distresses.

- **IRI Local Calibration Parameters**—Distress Specific = Global Calibration Values. The global SEE values will need to be used because the MEDPG software does not allow or permit those values to be changed without the source code.
- **Subgrade/Unbound Layer Local Calibration Parameter**—β_{s1}, = 0.5. A similar value has also been recommended for use from other calibration studies by individual state agencies. The SEE relationship based on the global calibration for the subgrade soils was not evaluated, because trenches were excluded from both demonstrations. The SEE relationship included in the MEPDG software for the unbound layers and soils was simply accepted as is.
- **Rut Depth Local Calibration Parameters**—The HMA local calibration parameters for rutting are mixture (PMA versus conventional mixtures) and volumetric property dependent. Previous studies have shown that PMA mixtures result in better performance than conventional type mixtures. The results from these demonstrations do not support that finding. It is recommended that forensic studies be implemented to determine why the PMA and Superpave mixtures have exhibited higher levels of rutting than conventional neat HMA mixtures. Until forensic investigations are completed, the overall average values for all mixes are recommended for use in day-to-day practice.
  - HMA Layers/Mixture Local Calibration Parameter, β_{r1}, = 1.5
  - HMA Layers/Mixture Local Calibration Parameter, β_{r3}, = 0.95
  - HMA Layers/Mixture Local Calibration Parameter, β_{r2}, = 1.0

The total rut depth SEE values derived from the local calibration process were lower than the SEE values included in the MEDPG Version 1.0 software for the total predicted rut depth (refer to Figures A2-13 and A-26). Thus, the SEE relationship for the HMA mixtures is recommended for use in day-to-day practice, as shown below.

$$SEE_{HMA} = 0.22\,(RD_{HMA})^{0.90} + 0.001 \qquad (A2\text{-}5)$$

- **Fatigue (Alligator) Cracking Local Calibration Parameters**—The HMA local calibration parameters for fatigue cracking are mixture (PMA versus conventional mixtures) and volumetric property dependent. Previous studies have shown that PMA mixtures result in better performance than conventional neat HMA mixtures. The results from this demonstration would not support that finding. It is recommended that forensic studies be implemented to determine why the PMA and Superpave mixtures have exhibited higher levels of fatigue cracking than the conventional neat HMA mixtures. Until forensic investigations are completed, the overall average values for all mixes are recommended for use in day-to-day practice.

- HMA Layers/Mixture Local Calibration Parameter, β_{f1}, = 1.5.
- HMA Layers/Mixture Local Calibration Parameters, β_{f2} and β_3 = 1.0.
- HMA Layers/Mixture Local Calibration Parameter, C_2, = 1.0;

Unity for this fatigue cracking parameter is believed to be low based on the test sections included in both demonstrations. However, there were too many sections with construction anomalies and too few with higher levels of fatigue cracking to determine a reliable estimate of this parameter. Thus, it is recommended that the global value of unity continue to be used.

The fatigue cracking SEE values derived from the local calibration were higher than the values derived from the global calibration for the greater areas of fatigue cracking. The reason for this reduced precision is a result of the inadequate bond hypothesis and anomalies reported during construction. The SEE fatigue cracking values derived from the PMS and LTPP data sets were different. Thus, the SEE relationship derived from the global calibration and included in the MEPDG software should continue to be used until more sections with higher amounts of fatigue cracking are included in the local calibration or forensic investigations are completed to confirm or reject the hypothesis.

- **HMA Thermal Cracking Local Calibration Parameter**, $\beta_{\tau3}$, = 2.0. The thermal cracking SEE values derived from the local calibration process were higher than the SEE values included in the MEDPG Version 1.0 software for both data sets (refer to Figures A2-13 and A2-26). One reason for the reduced model precision is probably related to the anomalies and harder asphalt properties reported during production. The SEE thermal cracking values derived from the PMS and LTPP data sets, however, were similar. Thus, the SEE relationship derived from the local calibration is recommended for use in day-to-day practice, as shown below.

$$SEE_{TC} = 0.22\,(TC)^{0.80} + 20 \qquad (A2\text{-}6)$$

A2.4 Attachments

A2.4.A Attachment A—Description of PMS Segments

A2.4.A.1 HMA Full-Depth New Construction/Reconstruction Projects

Project FDAC-C-2

The PMS ID for Project FDAC-C-2 is 067304017180. It is a 0.94-mi-long HMA full-depth reclamation project. This PMS segment is located in the northbound lanes of State Route 47 between mileposts 32.913 and 33.855 in Neosho County (KSDOT District 4). The HMA mixture placed along this project includes conventional, neat asphalt. KSDOT records show the new construction and maintenance and rehabilitation (M&R) performed on this section to date as the following:

- No record available for new construction.
- 1992 reconstruction (1-in. HMA + 8 in. recycled HMA + 6-in. lime subgrade + subgrade).
- 2001 overlay (1.6-in. HMA overlay).

The year of construction for this full-depth reclamation project was 1992 and performance data are available from 4/20/1993 (after the reconstruction) to 3/22/2001 (before the 2001 overlay was placed [refer to Table A2-1]). The International Roughness Index (IRI) measured along this segment of roadway in 1993 (85 in./mi) was selected as the initial IRI for this project.

Project FDAC-C-4

The PMS ID for Project FDAC-C-4 is 0223120008080. It is a 0.83-mi-long, full-depth HMA pavement with neat HMA mixtures. This PMS segment is located in the eastbound lanes of State Route 120 between mileposts 8.098 and 8.933 in Doniphan Country (KSDOT District 1). Information available from KSDOT records shows the new construction and M&R as the following:

- No record available for new construction.
- 1989 reconstruction (1.5-in. HMA + 6.5-in. HMA + subgrade).
- 1999 overlay (1.6-in. HMA overlay).

The year of reconstruction was 1989, and performance data are available from 3/28/1990 (after the reconstruction) to 3/21/1999 (before the overlay placement). The IRI measured for this segment of roadway in 1990 (110 in./mi) was selected as the initial IRI for this project.

Project FDAC-P-1

The PMS ID for Project FDAC-P-1 is 0552083020210. It is a 0.91-mi-long, full-depth HMA pavement with a PMA wearing course. This PMS segment is located in the northbound lanes of U.S. Route 83 between mileposts 144.713 and 145.625 in Logan Country (KSDOT District 3). Information available from KSDOT records shows the new construction and M&R as the following:

- No record available for new construction.
- 1999 reconstruction (1-in. HMA + 4-in. HMA + 8.7-in. HMA + 4-in. lime subgrade + subgrade).

The year of reconstruction was 1999, and performance data are available from 4/6/2000 (after the reconstruction) to 4/19/2006. The IRI measured for this segment of roadway in 2000 (36 in./mi) was selected as the initial IRI for this project.

Project FDAC-P-2

The PMS ID for Project FDAC-P-2 is 0552083026270. It is a 0.50-mi-long, full-depth HMA pavement with a PMA wearing course and binder layer. The project is located in the northbound lanes of U.S. Route 83 between mileposts 150.360 and 150.857 in Logan Country (KSDOT District 3). Information available from KSDOT records shows the new construction and M&R as the following:

- No record available for new construction.
- 1999 reconstruction (1-in. HMA + 4-in. HMA + 8.7-in. HMA + 4-in. lime subgrade + subgrade).

The year of reconstruction was 1999, and performance data are available from 4/6/2000 (after the reconstruction) to 4/19/2006. The IRI measured for this segment of roadway in 2000 (35 in./mi) was selected as the initial IRI for this project.

Project FDAC-P-3
The PMS ID of Project FDAC-P-3 is 0552083028290. It is a 0.47-mi-long, full-depth HMA pavement with a PMA wearing surface and binder layer. The project is located in the northbound lanes of U.S. Route 83 between mileposts 153.092 and 153.566 in Logan Country (KSDOT District 3).

Two pavement structures or cross sections were identified for this project, but the only difference between the two structures is the thickness of the HMA base layer. Section 1 has a HMA base thickness of 4.7 in. and is 0.34 mi in length, while Section 2 has a HMA base of 8.7 in. in thickness and is 0.13 mi in length. Since Section 1 is the longer segment, it was considered to represent the whole project and used in the local calibration process. Information available from KSDOT records shows the new construction and M&R as the following for Section 1:

- No record available for new construction.
- 1999 reconstruction (1.0-in. HMA + 4.0-in. HMA + 4.7-in. HMA + 3.5-in. lime modified subgrade + subgrade).

The year of reconstruction for both sections was 1999, and performance data are available from 4/6/2000 (after the reconstruction) to 4/19/2006. The IRI measured for this segment of roadway in 2000 (40.5 in./mi) was selected as the initial IRI for this project.

Project FDAC-P-4
The PMS ID of Project FDAC-P-4 is 0552083021220. It is a 0.45-mi-long, full-depth pavement project with PMA mixtures. This project is located in the northbound lanes of U.S. Route 83 between mileposts 145.625 and 146.075 in Logan Country (KSDOT District 3). Information available from KSDOT records shows the new construction and M&R as the following:

- No record available for new construction.
- 1999 reconstruction (1-in. HMA + 4-in. HMA + 8.7-in. HMA + 4-in. lime subgrade + subgrade).

The year of reconstruction was 1999, and performance data are available from 4/6/2000 (after the reconstruction) to 4/19/2006. The IRI measured for this segment of roadway in 2000 (39.0 in./mi) was selected as the initial IRI for the project.

Project FDAC-P-5
The PMS ID of Project FDAC-P-5 is 0552083023240. It is a 0.38-mi-long, full-depth PMA pavement project. This PMS segment is located in the northbound lanes of U.S. Route 83 between mileposts 148.185 and 148.566 in Logan Country (KSDOT District 3). Information available from KSDOT records shows the new construction and M&R as the following:

- No record available for new construction.
- 1999 reconstruction (1-in. HMA + 4-in. HMA + 8.7-in. HMA + 4-in. lime subgrade + subgrade).

The year of reconstruction was 1999, and performance data are available from 4/6/2000 (after the reconstruction) to 4/19/2006. The IRI measured for this segment in 2000 (37.0 in./mi) was selected as the initial IRI for this project.

Project FDAC-S-1

The PMS ID for Project FDAC-S-1 is 0372054012120. It is a 0.67-mi-long, full-depth HMA project and all layers were designed in accordance with Superpave. The project is located in the eastbound lanes of U.S. Route 54 between mileposts 282.514 and 283.193 in Greenwood Country (KSDOT District 4). Information available from KSDOT records shows the new construction and M&R as the following:

- No record available for new construction.
- 1998 reconstruction (1-in. HMA + 15-in. HMA + 6-in. lime subgrade + subgrade).

The year of reconstruction was 1998, and performance data are available from 4/21/1999 (after the reconstruction) to 3/29/2006. The IRI measured for this roadway segment in 1999 (44.0 in./mi) was selected as the initial IRI for the project.

Project FDAC-S-3

The PMS ID for Project FDAC-S-3 is 0702075007080. It is a 0.84-mi-long, full-depth HMA pavement with Superpave mixtures. This PMS segment is located in the northbound lanes of U.S. Route 75 between mileposts 120.227 and 121.068 in Osage Country (KSDOT District 1). Information available from KSDOT records shows the new construction and M&R as the following:

- No record available for new construction.
- 1998 reconstruction (1-in. HMA + 4-in. HMA + 6-in. HMA + 6-in. lime subgrade + subgrade).

The year of reconstruction was 1998, and performance data are available from 3/15/1999 (after the reconstruction) to 2/27/2006. The IRI measured along this roadway segment in 1999 (55.0 in./mi) was selected as the initial IRI for the project.

Project FDAC-S-4

The PMS ID for Project FDAC-S-4 is 067304017180. It is a 0.81-mi-long, full-depth HMA pavement with Superpave mixtures. This PMS segment is located in the northbound lanes of U.S. Route 75 between mileposts 32.679 and 33.493 in Montgomery Country (KSDOT District 4). Information available from KSDOT records shows the new construction and M&R as the following:

- No record available for new construction.
- 1998 reconstruction (1-in. HMA + 12-in. HMA + 6-in. lime subgrade + subgrade).

The year of reconstruction was 1998, and performance data are available from 4/14/1999 (after the reconstruction) to 3/13/2006. The IRI measured for this segment of roadway in 1999 (54.0 in./mi) was selected as the initial IRI for the project.

Project FDAC-S-5

The PMS ID for Project FDAC-S-5 is 0702075008090. It is a 4.21-mi-long, full-depth HMA pavement with Superpave mixtures. This PMS segment is located in the northbound lanes of U.S. Route 75 between mileposts 121.068 and 125.275 in Osage Country (KSDOT District 1). Information available from KSDOT records shows the new construction and M&R as the following:

- No record available for new construction.
- 1998 reconstruction (1-in. HMA + 4-in. HMA + 6-in. HMA + 6-in. lime subgrade + subgrade).

The year of reconstruction was 1998, and performance data are available from 3/15/1999 (after the reconstruction) to 2/27/2006. The IRI measured for this segment of roadway in 1999 (40.0 in./mi) was selected as the initial IRI for this project.

A2.4.A.2 HMA Overlay of Flexible Pavement Projects

HMA-Overlay Project: HMA_HMA-C-1

The PMS ID for Project HMA_HMA-C-1 is 0702075008090. It is a 1.39-mi-long, HMA-Overlay Project that is located in the northbound lanes of State Route 177 between mileposts 7.414 and 8.807 in Butler Country (KSDOT District 5). The overlay consists of a conventional neat HMA mixture. Information available from KSDOT records shows the new construction and M&R as the following:

- 1979 new construction (1-in. HMA + 7-in. HMA + 8-in. lime subgrade + subgrade).
- 1989 overlay ($^3/_4$-in. overlay + 1-in. recycled HMA).
- 1997 overlay (3-in. HMA overlay).

The year for the placement of the second overlay was 1997. This second overlay was the one used in the local calibration process. Performance data are available from 4/16/1998 (after the placement of the second overlay) to 2/27/2006. The IRI measured for this segment of roadway in 1998 (70.5 in./mi) was selected as the initial IRI for this project.

HMA-Overlay Project: HMA_HMA-C-2

The PMS ID for Project HMA_HMA-C-2 is 0083177006070. It is a 2.16-mi-long, HMA-Overlay Project located in the northbound lanes of State 177 between mileposts 5.255 and 7.414 in Butler Country (KSDOT District 5). The overlay consists of a conventional neat HMA mixture. Information available from KSDOT records shows the new construction and M&R as the following:

- 1979 new construction (1-in. HMA + 5-in. HMA + 8-in. lime subgrade+ subgrade).
- 1989 overlay ($^3/_4$-in.-overlay + 1 in. recycled HMA).
- 1997 overlay (3-in. HMA overlay).

The year for the placement of the second overlay was 1997. This second overlay was the one used in the local calibration process. Performance data are available from 4/16/1998 (after the placement of the second overlay) to 3/28/2006. The IRI measured for this segment of roadway in 1998 (61.5 in./mi) was selected as the initial IRI for this project.

HMA-Overlay Project: HMA_HMA-C-4

The PMS ID for Project HMA_HMA-C-4 is 0132160005060. It is a 0.29-mi-long, HMA-Overlay Project located in the eastbound lanes of U.S. Route 160 between mileposts 135.261 and 135.550 in Clark Country (KSDOT District 6). The overlay consists of a conventional neat HMA mixture. Information available from KSDOT records shows the new construction and M&R as the following:

- 1964 new construction (7.5-in. HMA + subgrade).
- 1996 overlay (1-in. + 4.5-in. HMA overlay).

The year for the placement of the overlay was 1996, and performance data are available from 3/20/1997 (after the placement of the overlay) to 4/25/2006. The IRI measured for this segment of roadway in 1997 (37 in./mi) was selected as the initial IRI for the HMA overlay.

HMA-Overlay Project: HMA_HMA-P-3

The PMS ID for Project HMA_HMA-P-3 is 0233010007080. It is a 1.10-mi-long, HMA-Overlay Project located in the eastbound lanes of State Route 10 between mileposts 135.261 and 135.550 in Douglas Country (KSDOT District 1). The overlay consists of a PMA mixture. Information available from KSDOT records shows the new construction and M&R as the following:

- No record available for new construction.
- 1996 reconstruction (1-in. HMA + 8-in. HMA + 6-in. lime subgrade + subgrade).
- 2000 overlay (1.6-in. + 2.0-in. HMA overlay).

The year for the placement of the overlay was 2000, and performance data are available from 03/12/2001 (after the placement of the overlay) to 03/02/2006. The measured IRI for this project in 2001 (49.5 in./mi) is selected as the initial IRI.

HMA-Overlay Project: HMA_HMA-S-3

The PMS ID for Project HMA_HMA-S-3 is 0052281018200. The project contains two pavement structure sections with lengths of 0.04 miles (Section 1) and 0.78 miles (Section 2). Since Section 2 is the predominant section in length, only Section 2 was used in the local calibration process. This rehabilitation project is located in the northbound lanes of U.S. 281 between mileposts 115.539 and 116.315 in Barton Country (KSDOT District 5). The HMA overlay for this segment is a Superpave mixture. Information available from KSDOT records shows the new construction and M&R as the following:

- 1965 new construction (1.5-in. + 3.0-in. HMA + subgrade).
- 1999 overlay (1.5-in. HMA + 4.0-in. recycled HMA).
- 2005 overlay (1-in. overlay Superpave).

The year for the placement of the first overlay was 1999, and performance data are available from 04/13/2000 (after the placement of the first overlay) to 04/05/2005 (before the placement of the second overlay). The IRI measured for this segment of roadway in 2000 (53.5 in./mi) was selected as the initial IRI for the Superpave overlay.

A2.4.B Attachment B—Plots of Time-History Performance Data

The following are plots of time-history performance data that provide performance data over time for all of the Kansas PMS segments. These time-history plots were used to determine if any anomalies or outliers were present in the data. The first section provides the plots of each performance indicator for the full-depth HMA pavements, while the second section is for the HMA overlays.

A2.4.B.1 Full-Depth HMA, New Construction

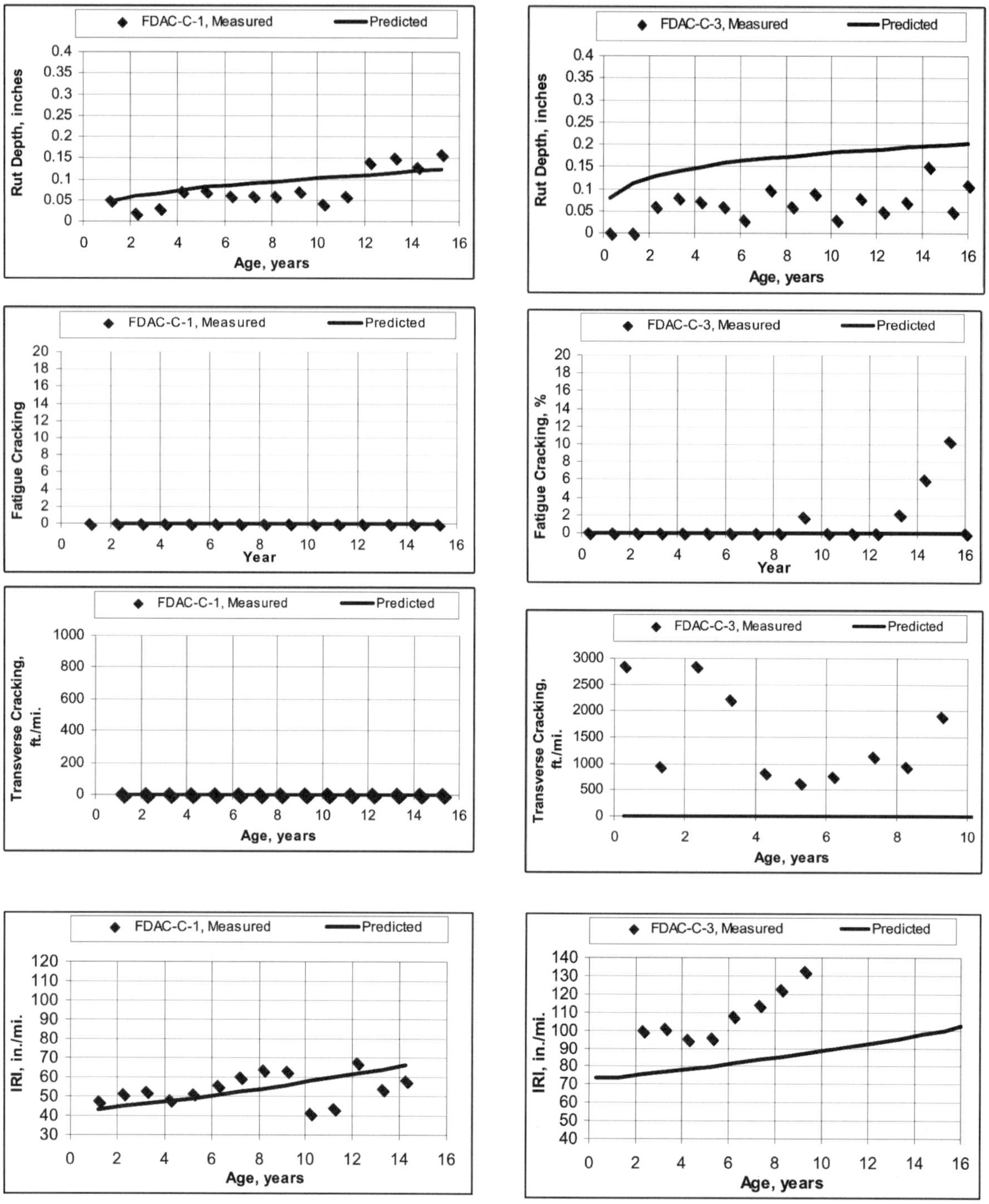

Figure A2.4.B.1-1. Full-Depth HMA, New Construction

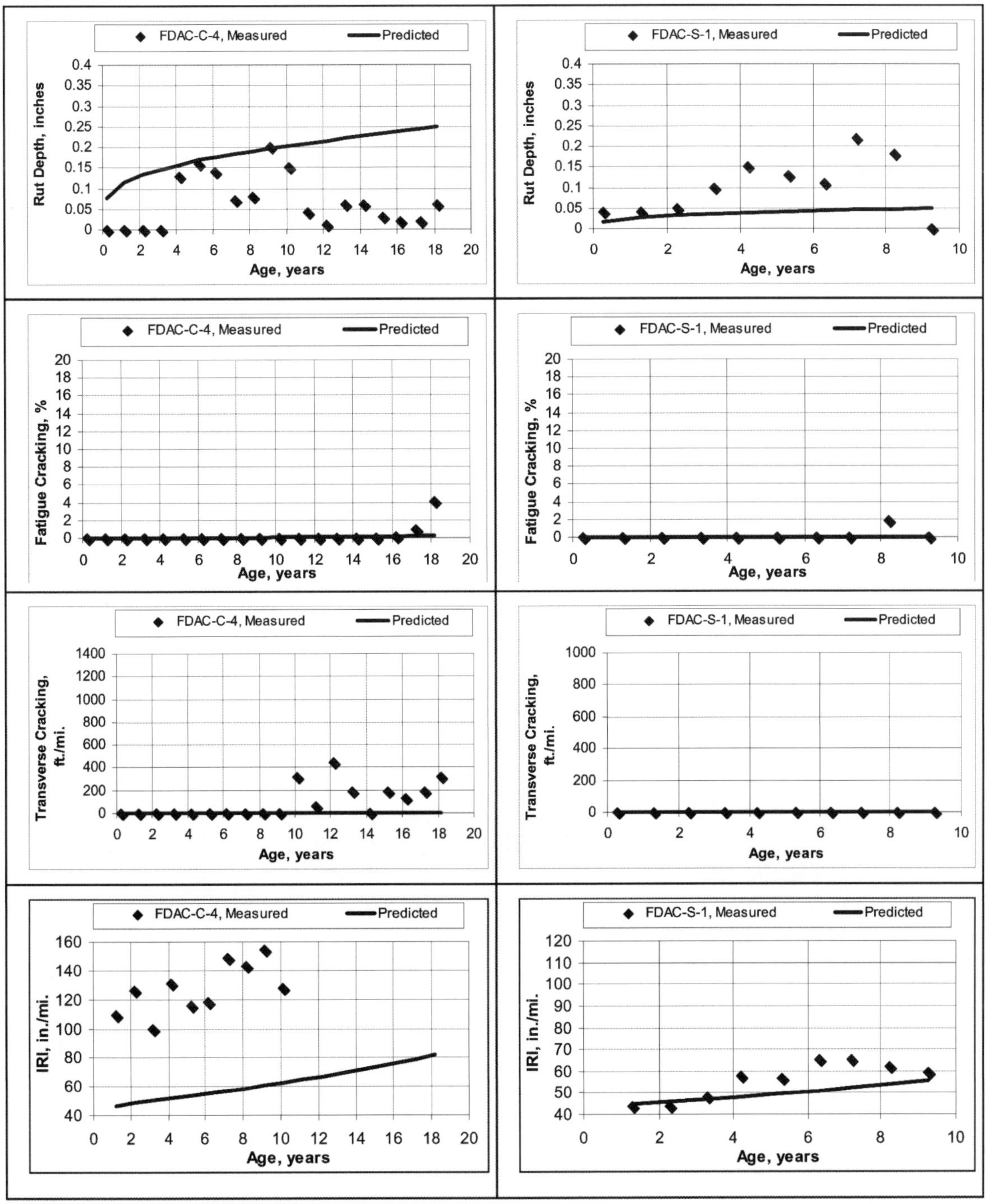

Figure A2.4.B.1-1.—*Continued*

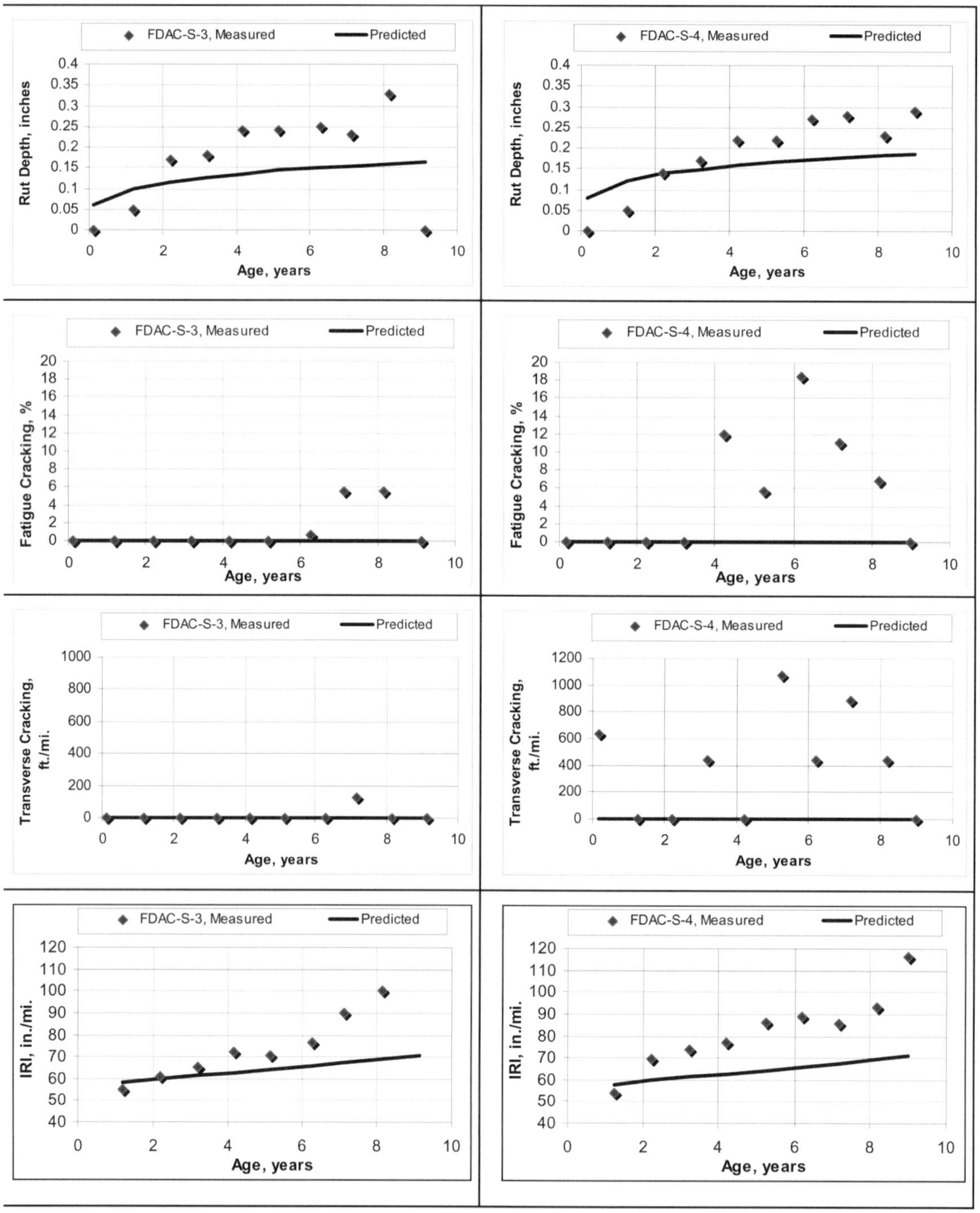

Figure A2.4.B.1-1.—*Continued*

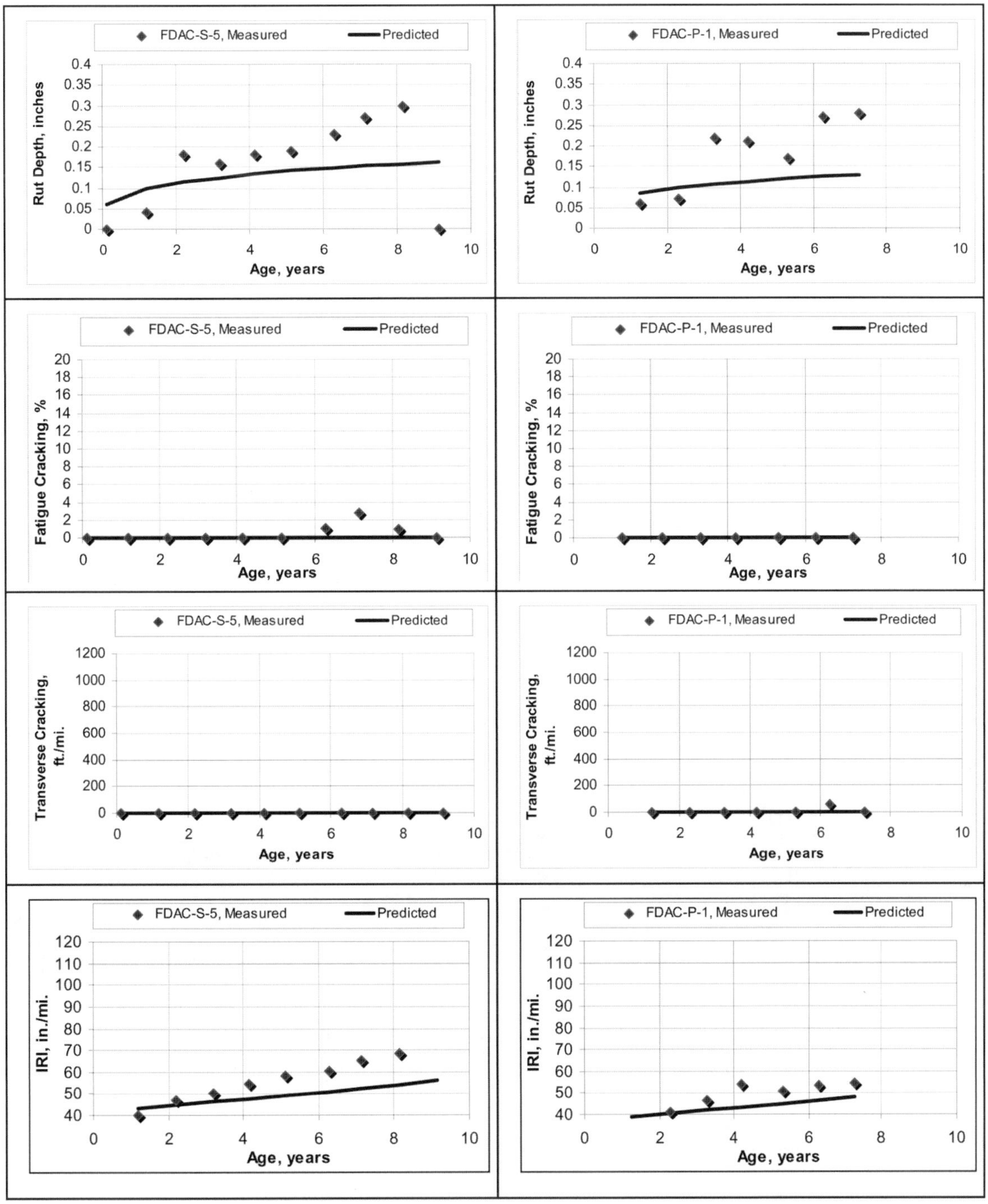

Figure A2.4.B.1-1.—*Continued*

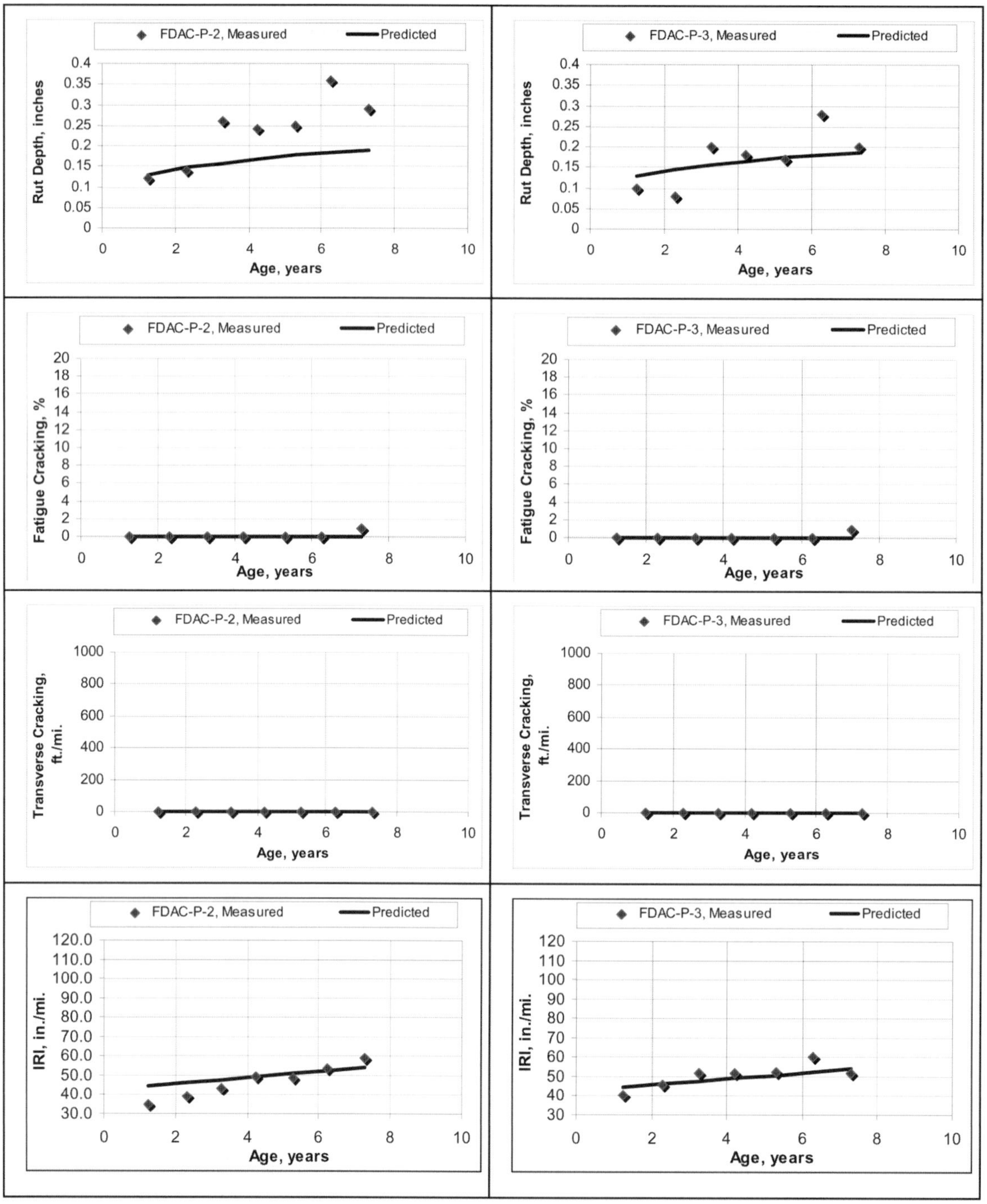

Figure A2.4.B.1-1.—*Continued*

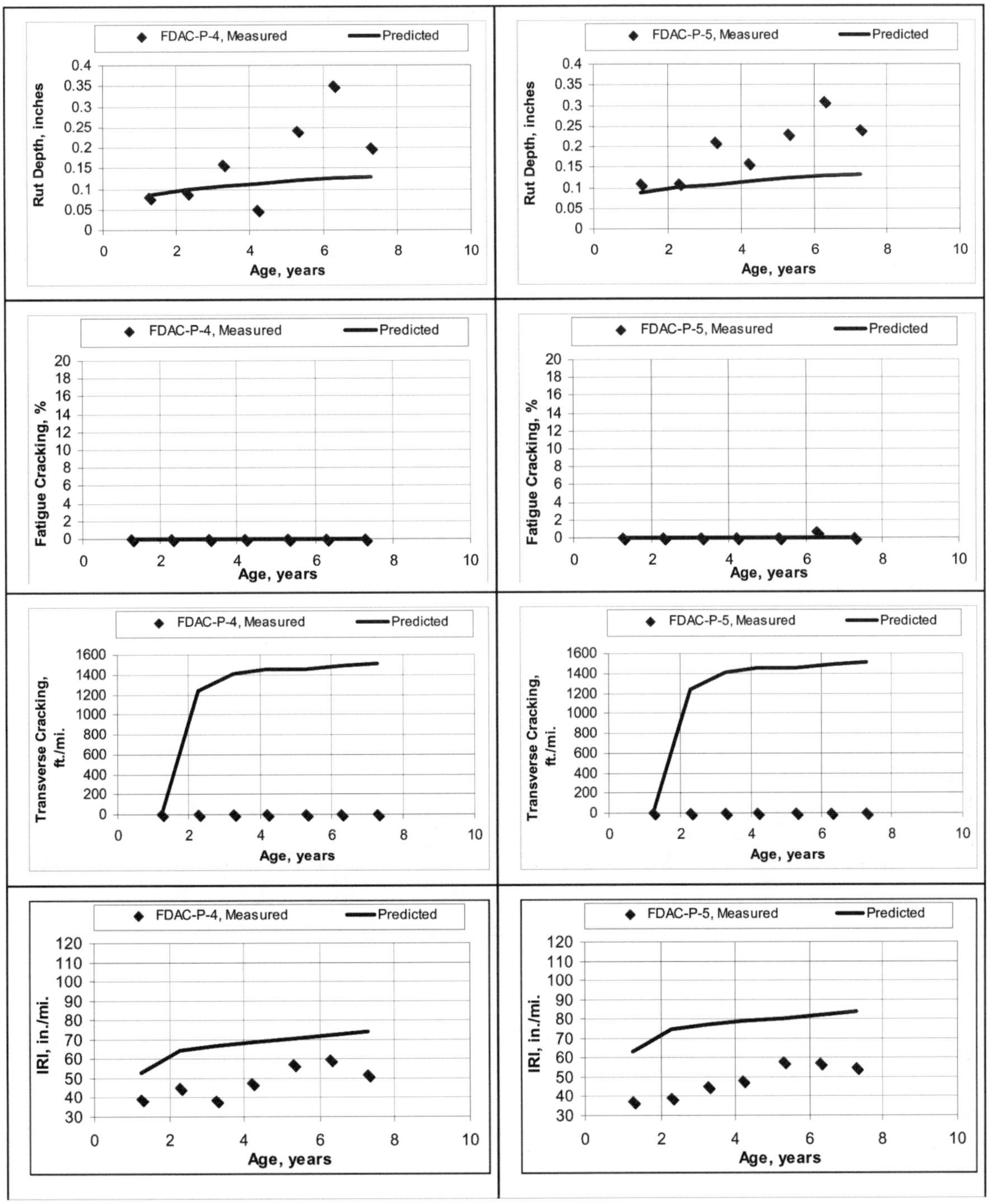

Figure A2.4.B.1-1.—*Continued*

A2.4.B.2 HMA Overlays of Flexible Pavements, Rehabilitation

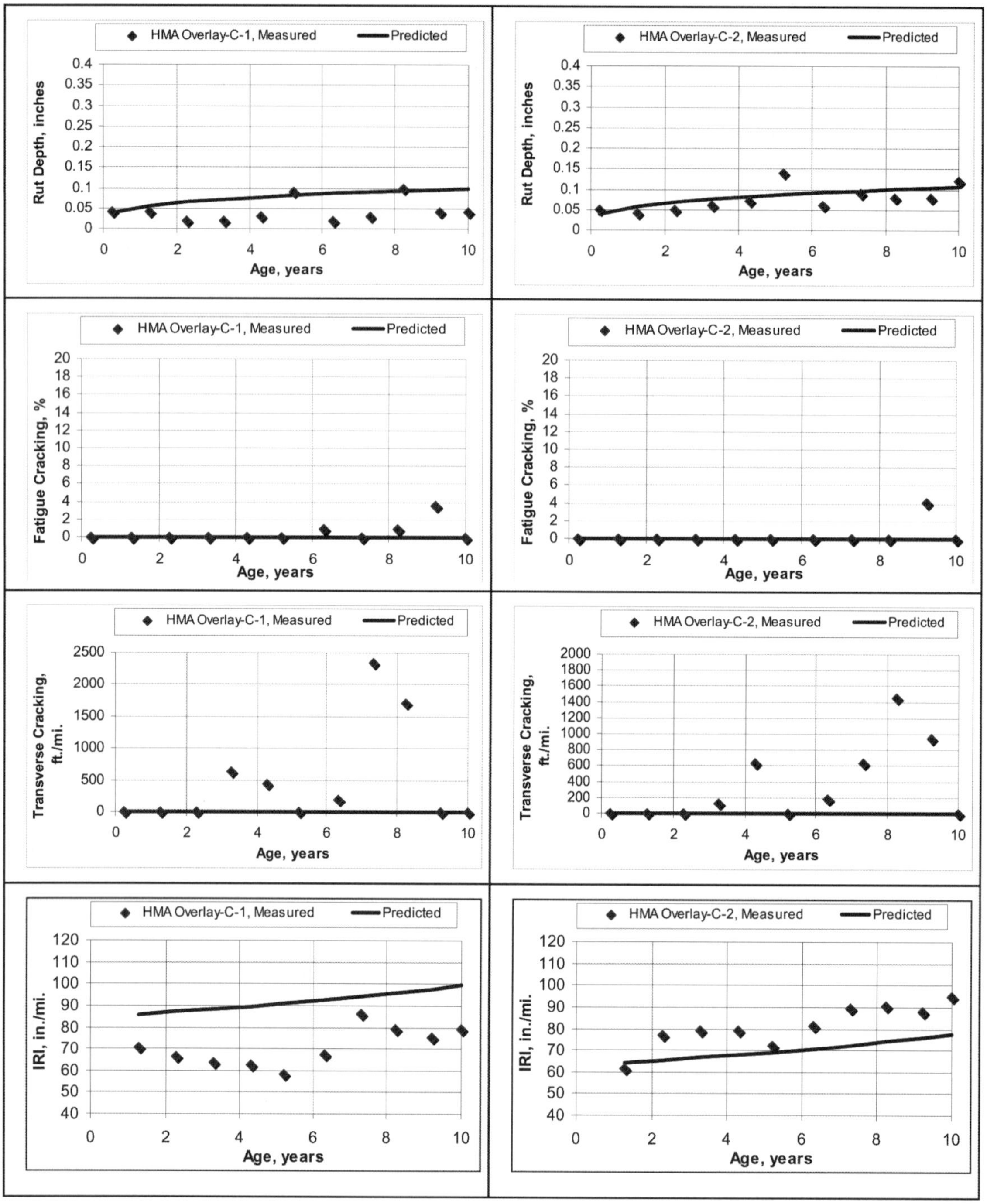

Figure A2.4.B.2-1. HMA Overlays of Flexible Pavements, Rehabilitation

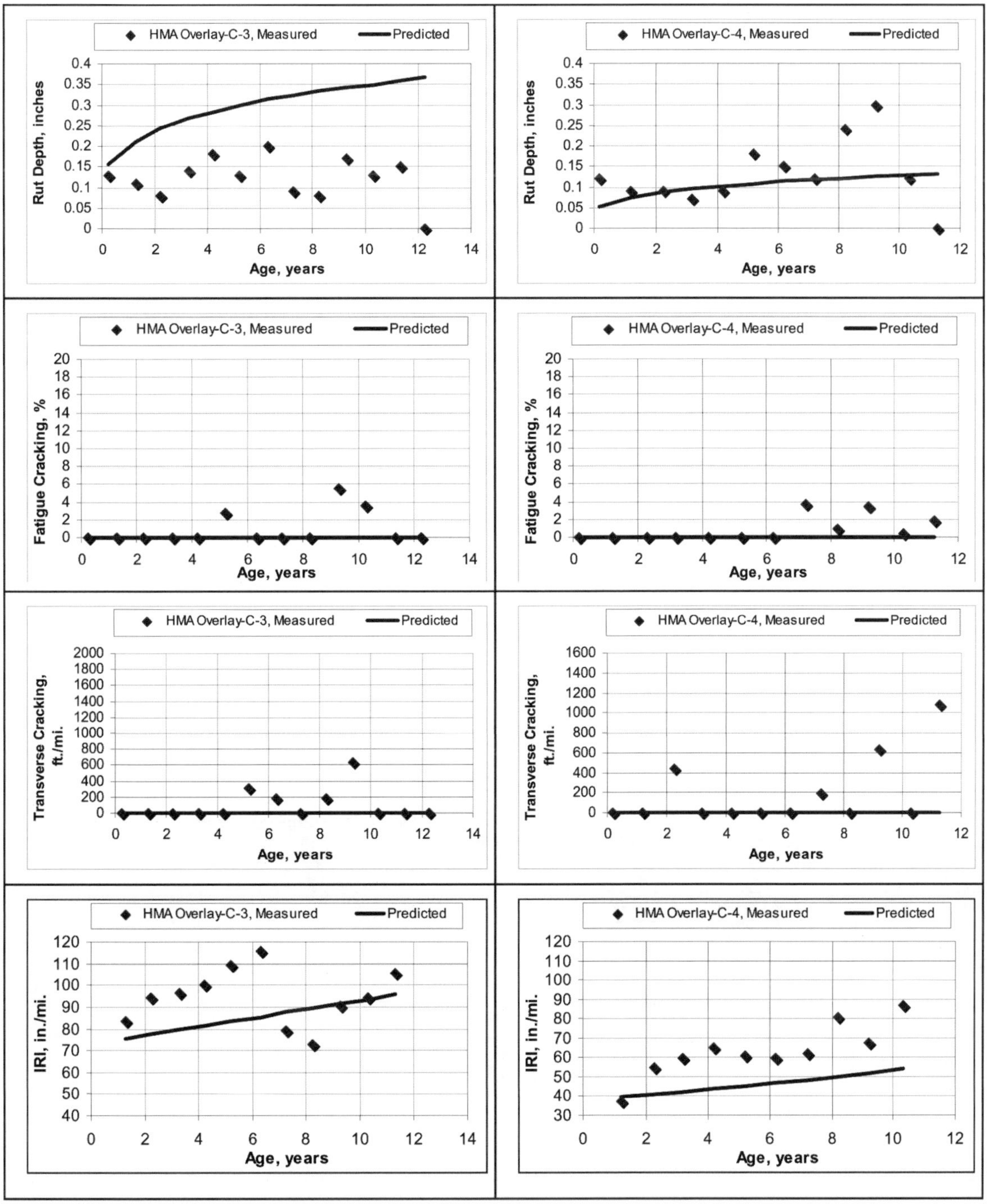

Figure A2.4.B.2-1.—*Continued*

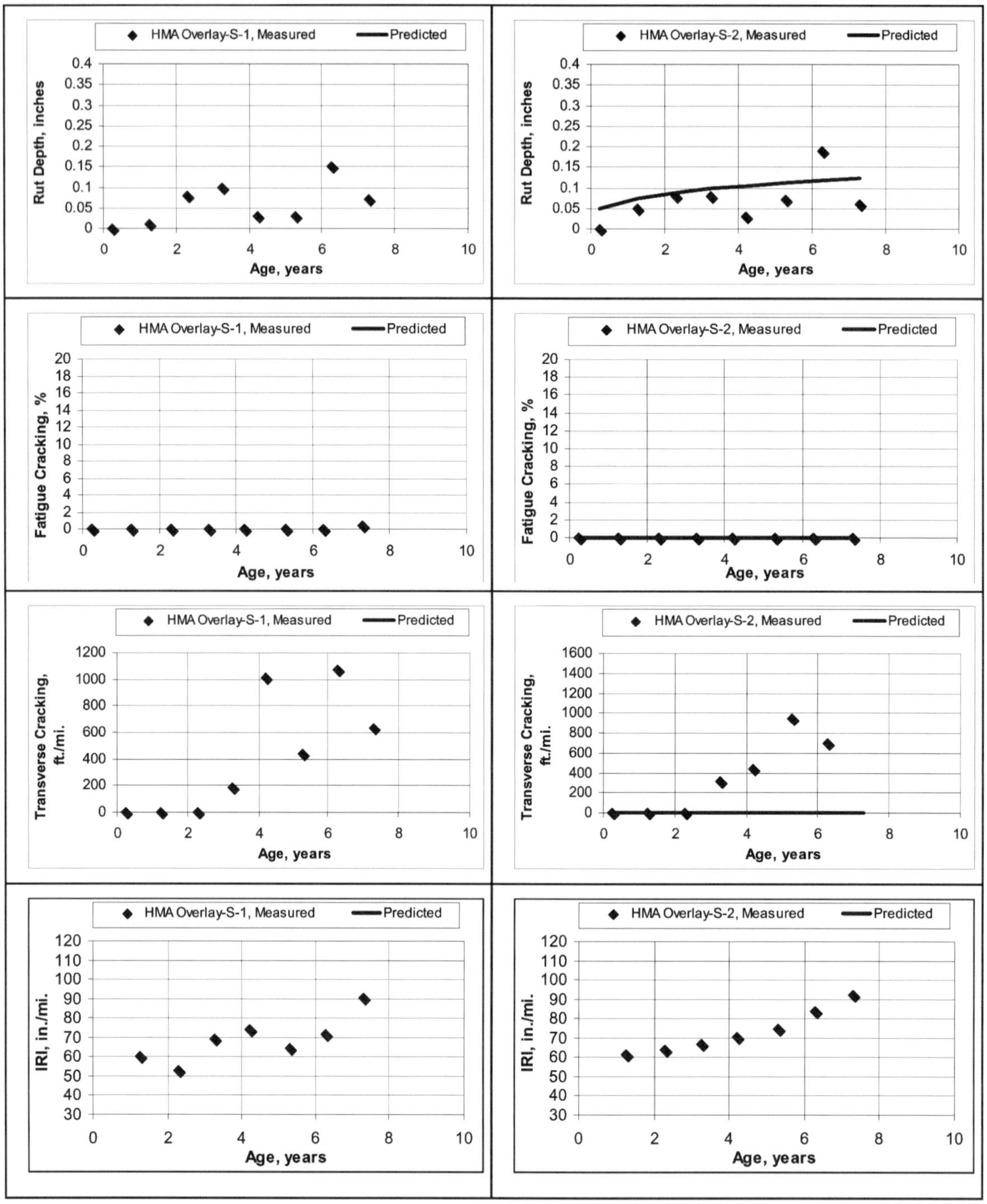

Figure A2.4.B.2-1.—*Continued*

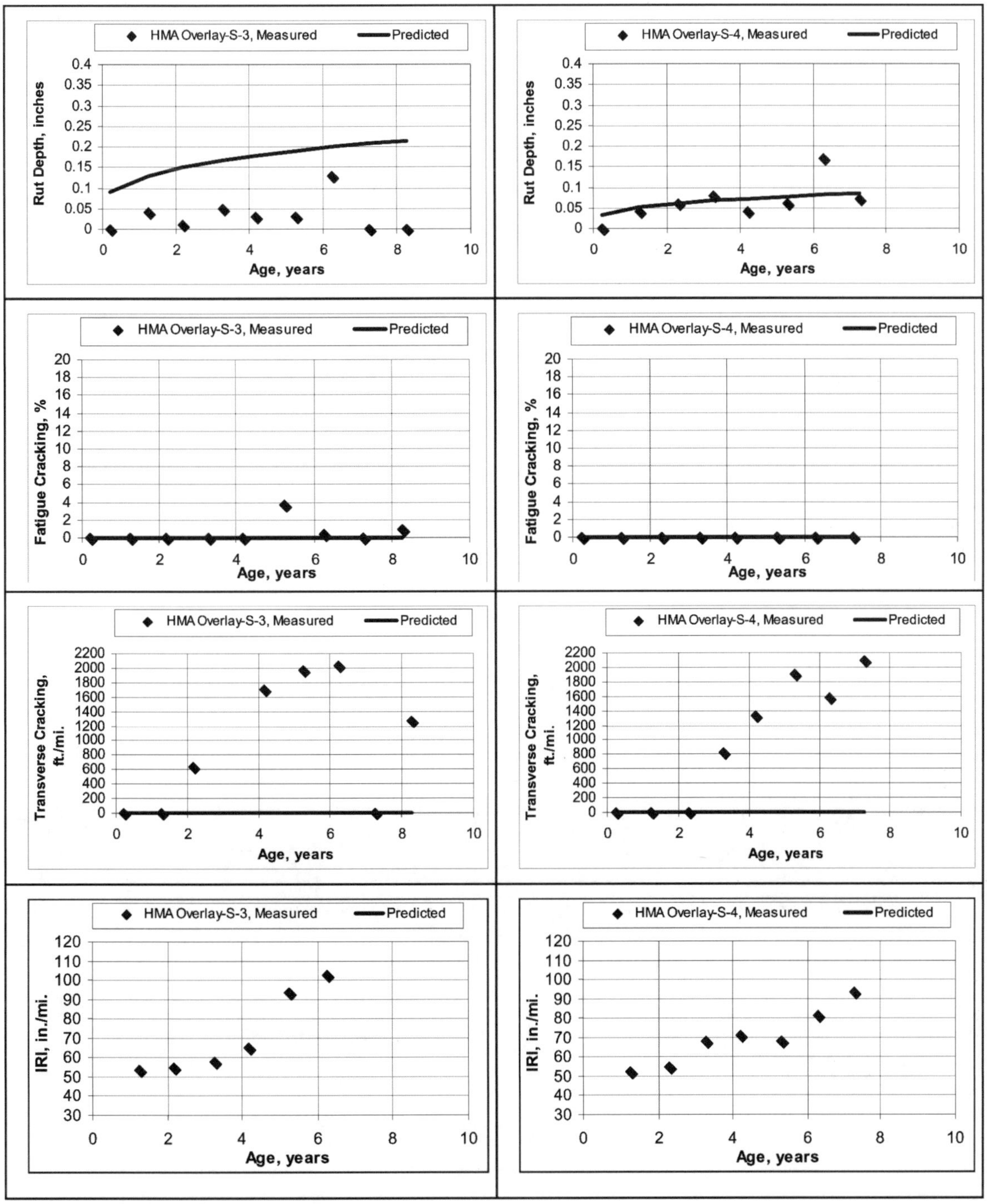

Figure A2.4.B.2-1.—*Continued*

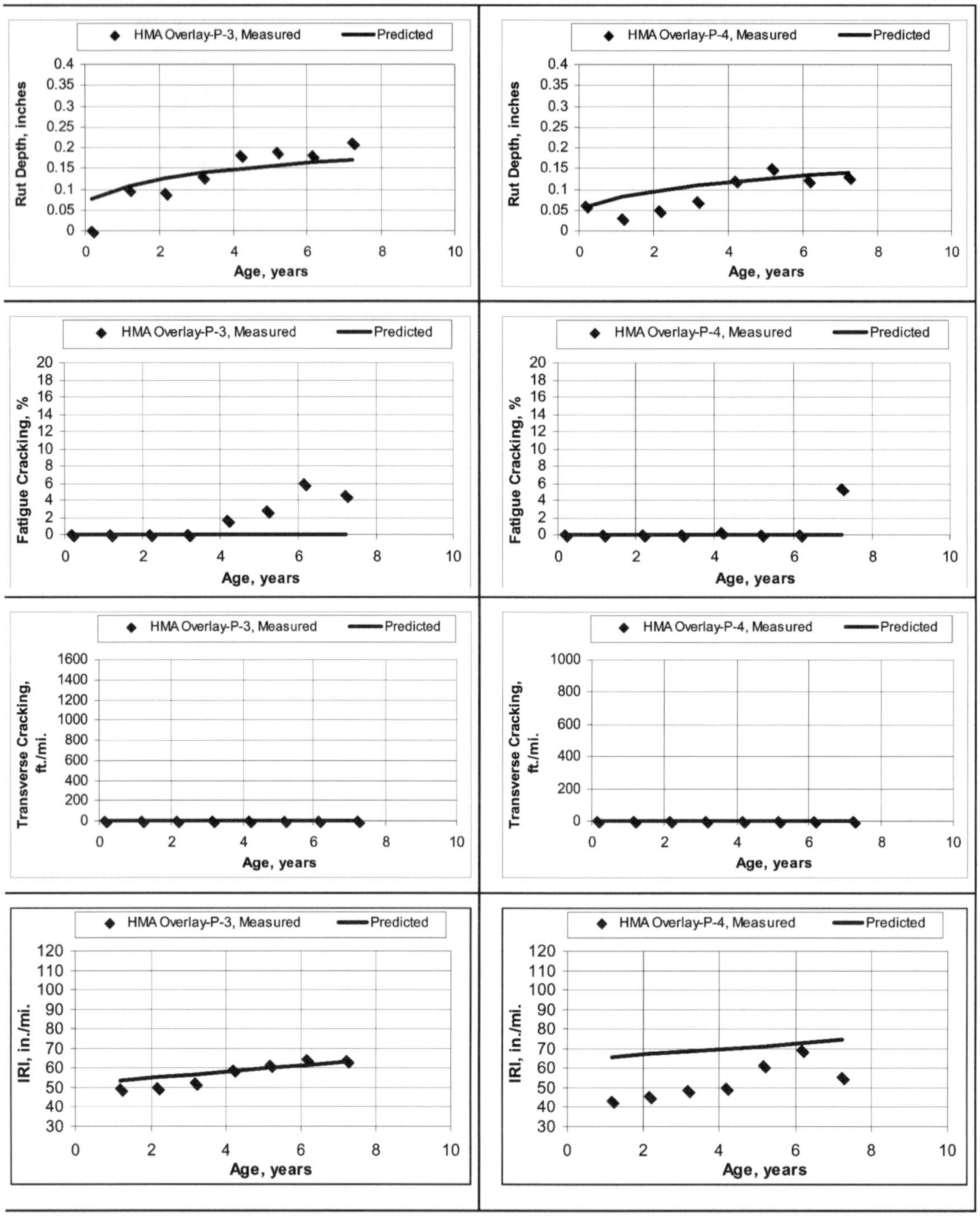

Figure A2.4.B.2-1.—*Continued*

A2.4.C Attachment C—Description of LTPP Projects

A2.4.C.1 Full-Depth and Conventional New Construction—LTPP SPS-1 Projects

Four LTPP SPS-1 projects were selected for the local calibration demonstration: Kansas, Iowa, Nebraska, and Oklahoma. None of the sections with ATPB layers were used for the demonstration. The construction reports were reviewed to identify items that occurred during construction that could have an impact on the performance of the sections, which are listed below.

- **Kansas**—Extensive rain delays occurred during construction. Excess moisture was reported in the unbound layers, specifically in the subbase. The high moisture content made compaction difficult. Some fly ash was added to the subbase material to facilitate compaction.

- **Iowa**—Rain delays occurred during construction. The unbound surfaced were dried and reworked prior to placing the first HMA layer. Some of the HMA mixtures were produced at high temperatures. These high temperatures were believed to be a result of higher water contents in the aggregate stockpiles for drying the aggregate during production. The viscosities reported during construction were high, suggesting that the asphalt had been hardened during construction.

- **Nebraska**—Rain delays occurred during construction. No problems were noted during construction, but the moisture contents of the unbound layers were high. Three of the sections were placed over culverts. The report did not identify the specific sections placed over the culverts.

- **Oklahoma**—The earthwork along this segment of roadway had been performed 10 years prior to construction. Cut and fill arrears occur along the segment of roadway with the SPS-1 project. No other construction deviations or problems were noted in the construction report.

A2.4.C.2 HMA Overlays of Flexible Pavement—LTPP SPS-5 Projects

Four LTPP SPS-5 projects were selected for the local calibration demonstration: Colorado, Missouri, Oklahoma, and Texas. The construction reports were reviewed to identify items that occurred during construction that could have an impact on the performance of the sections, which are listed below.

- **Colorado**—Asphalt viscosities reported during construction exceed 6,000 poises, and were greater than the viscosities measured on the existing HMA layers. A level-up layer was placed along all test sections, including the control section. The thickness of this level-up layer varied between the test sections.

- **Missouri**—No deviations or problems were reported during construction.

- **Oklahoma**—The first batch of RAP used in the HMA mixtures was reported to have high asphalt content. Asphalt viscosities exceeding 9,000 poises were reported during construction. The high asphalt viscosities suggest that the asphalt could have been hardened during construction.

- **Texas**—Asphalt viscosities exceeding 350,000 poises were reported during construction. These high viscosities suggest that a higher amount of filler had been used and/or that the asphalt had been severely hardened during construction. Problems with the plant were noted in the construction report. Air voids were low and the asphalt contents were highly variable during construction. The construction report also notes problems with the mixture design. Rain delays occurred during construction.

A2.4.D Attachment D—Plots of Time-History Performance Data for the LTPP SPS Projects

The following are plots of time-history performance data that provide performance data over time for selected LTPP SPS projects in and near Kansas. These time-history plots were used to determine if any anomalies or outliers were present in the data. The first section provides the plots of each performance indicator and the residual error using the global calibration values for the HMA pavements, while the second section is for the HMA overlays.

2.4.D.1 Full-Depth and Conventional HMA, New Construction

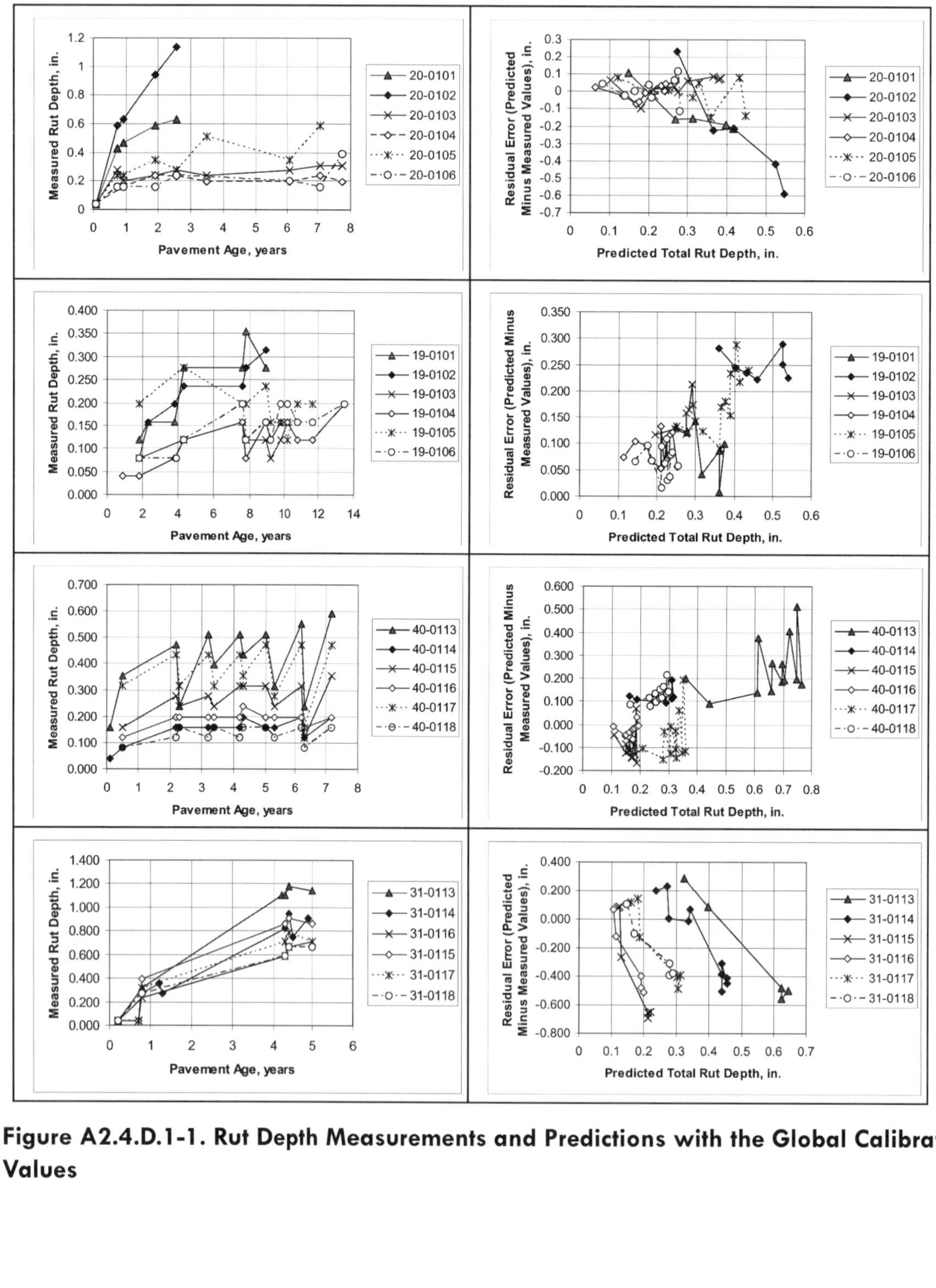

Figure A2.4.D.1-1. Rut Depth Measurements and Predictions with the Global Calibration Values

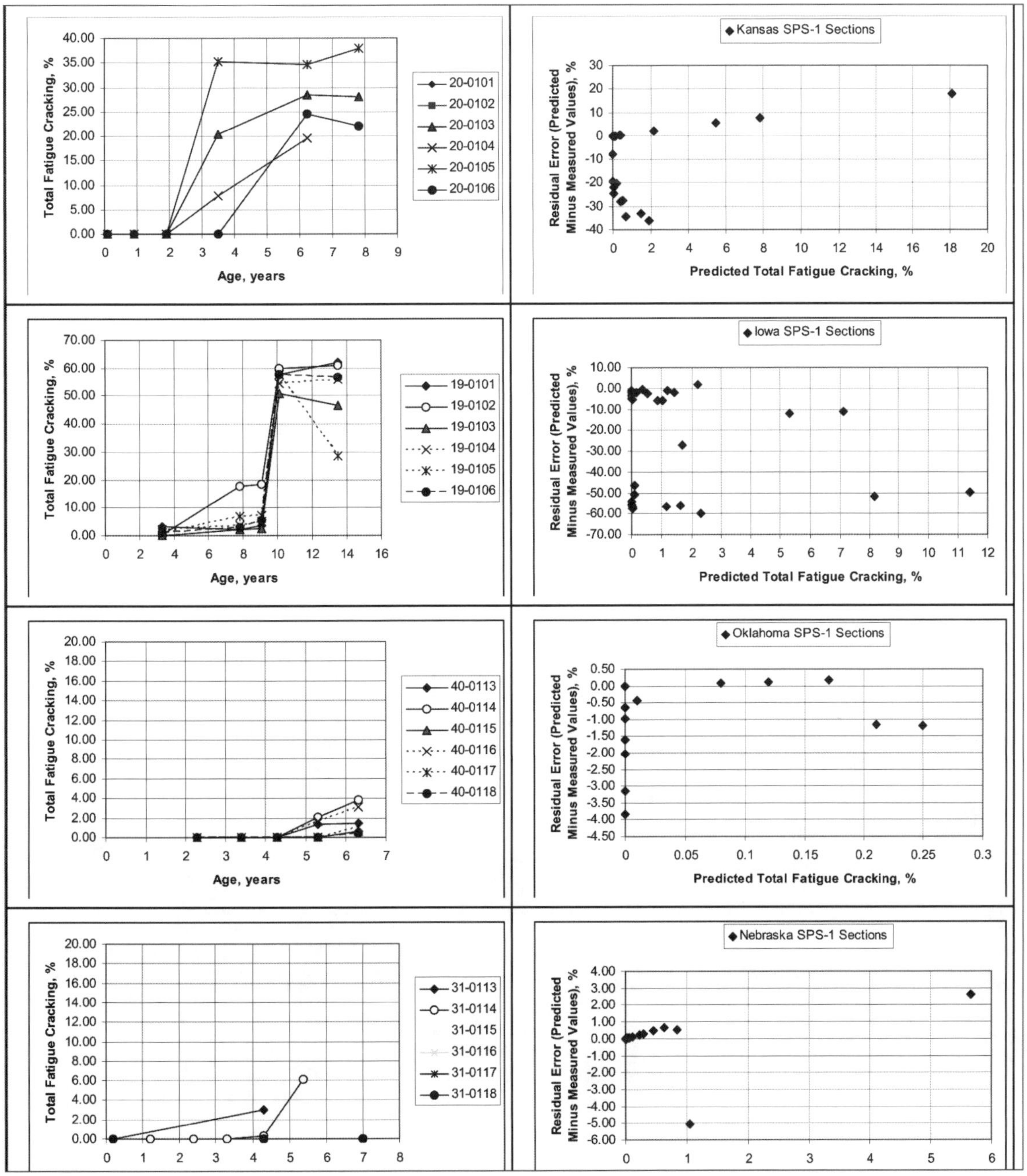

Figure A2.4.D.1-2. Fatigue Cracking Measurements and Predictions with the Global Calibration Values

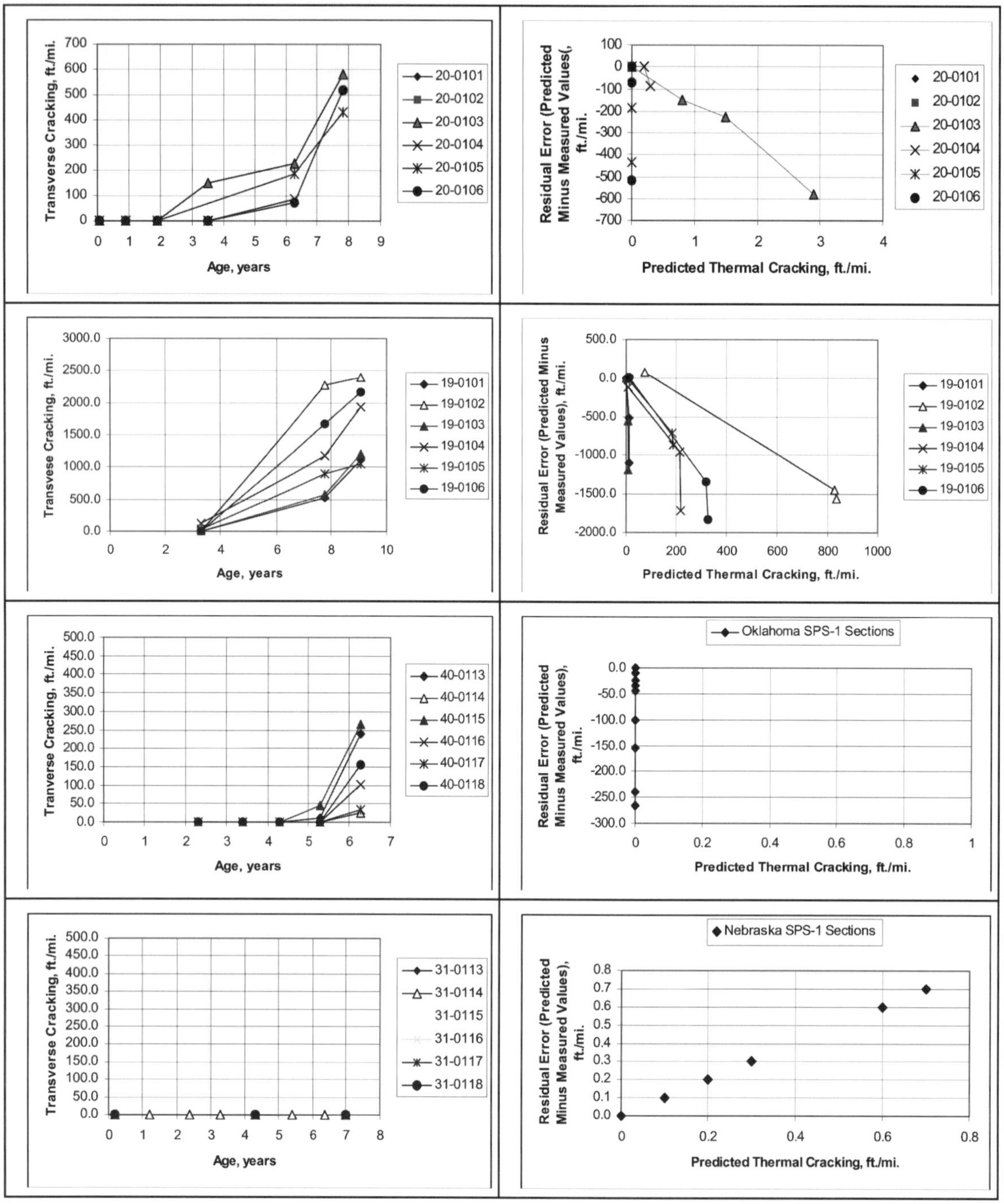

Figure A2.4.D.1-3. Transverse Cracking Measurements and Thermal Cracking Predictions with the Global Calibration Values

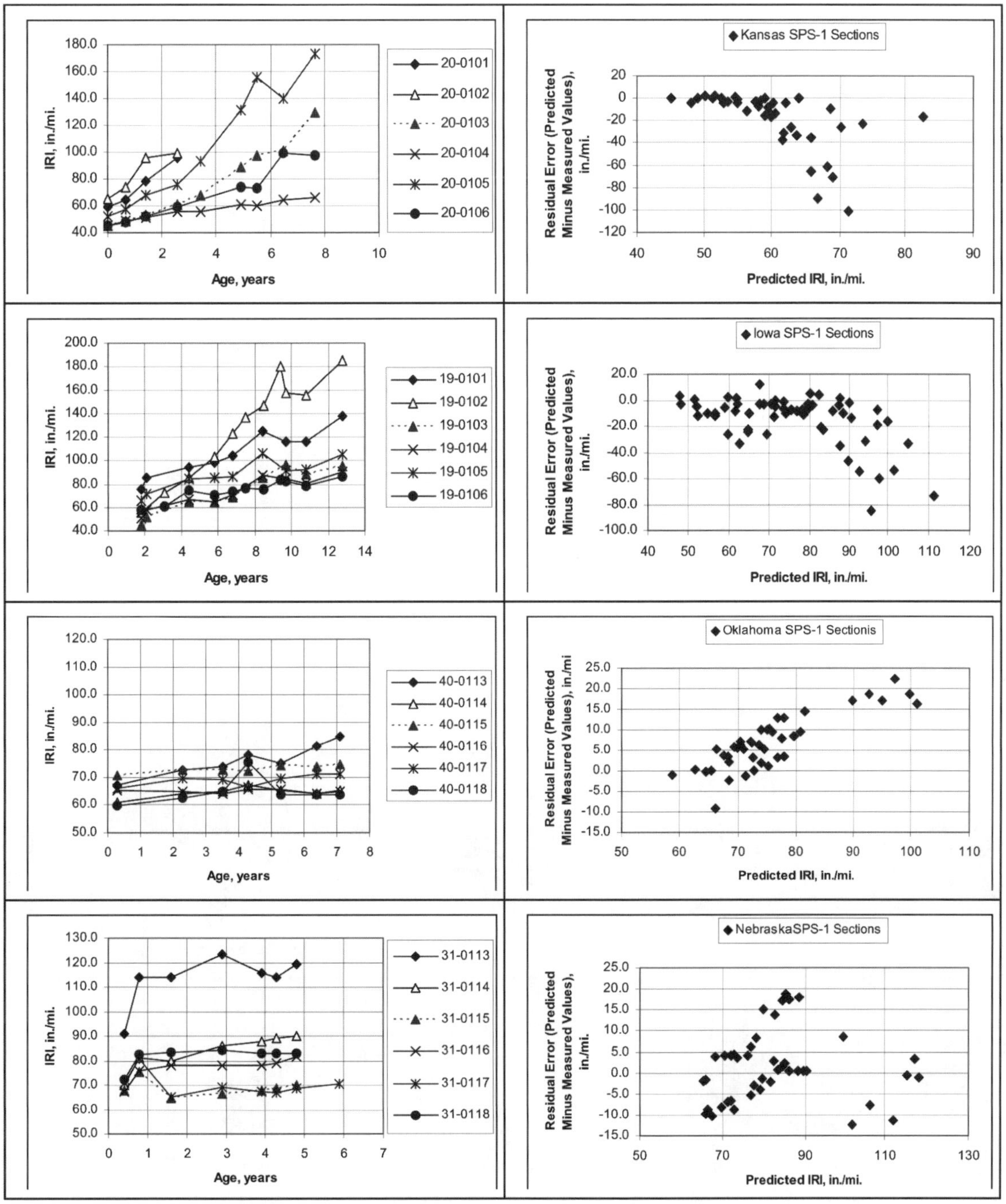

Figure A2.4.D.1-4. IRI Measurements and Predictions with the Global Calibration Values

A2.4.D.2 HMA Overlays of Flexible Pavements, Rehabilitation

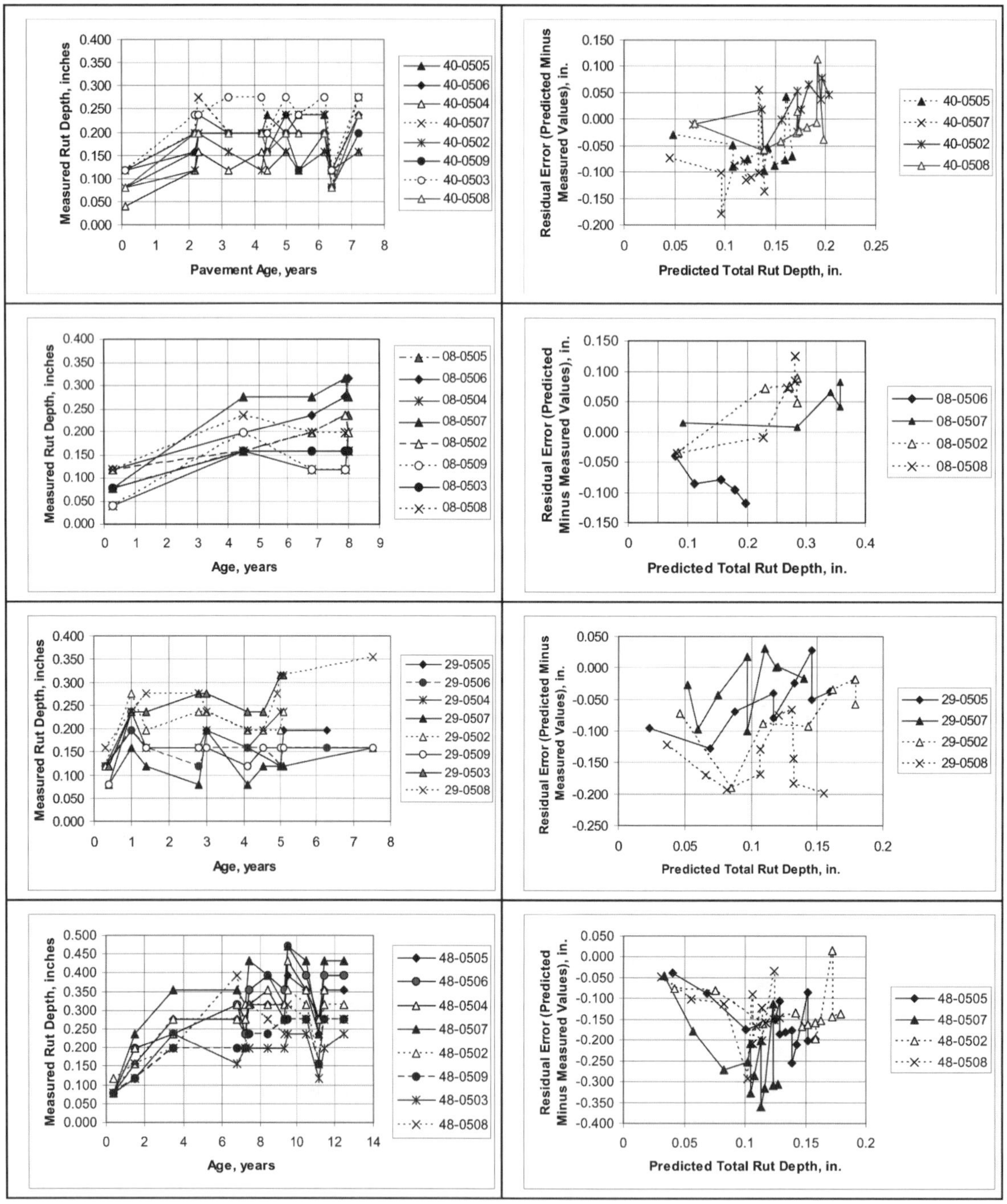

Figure A2.4.D.2-1. Rut Depth Measurements and Predictions with the Global Calibration Values

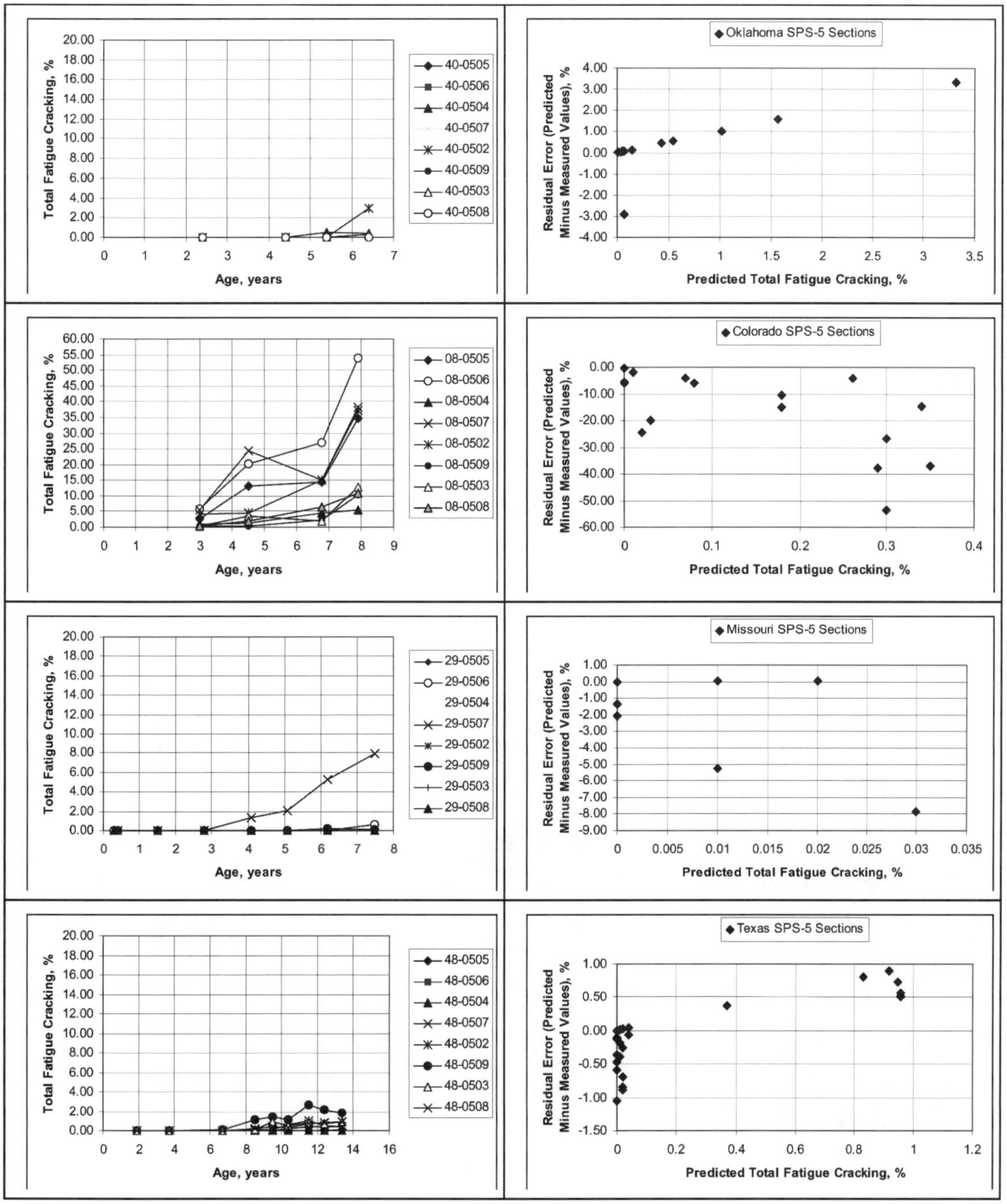

Figure A2.4.D.2-2. Fatigue Cracking Measurements and Predictions with the Global Calibration Values

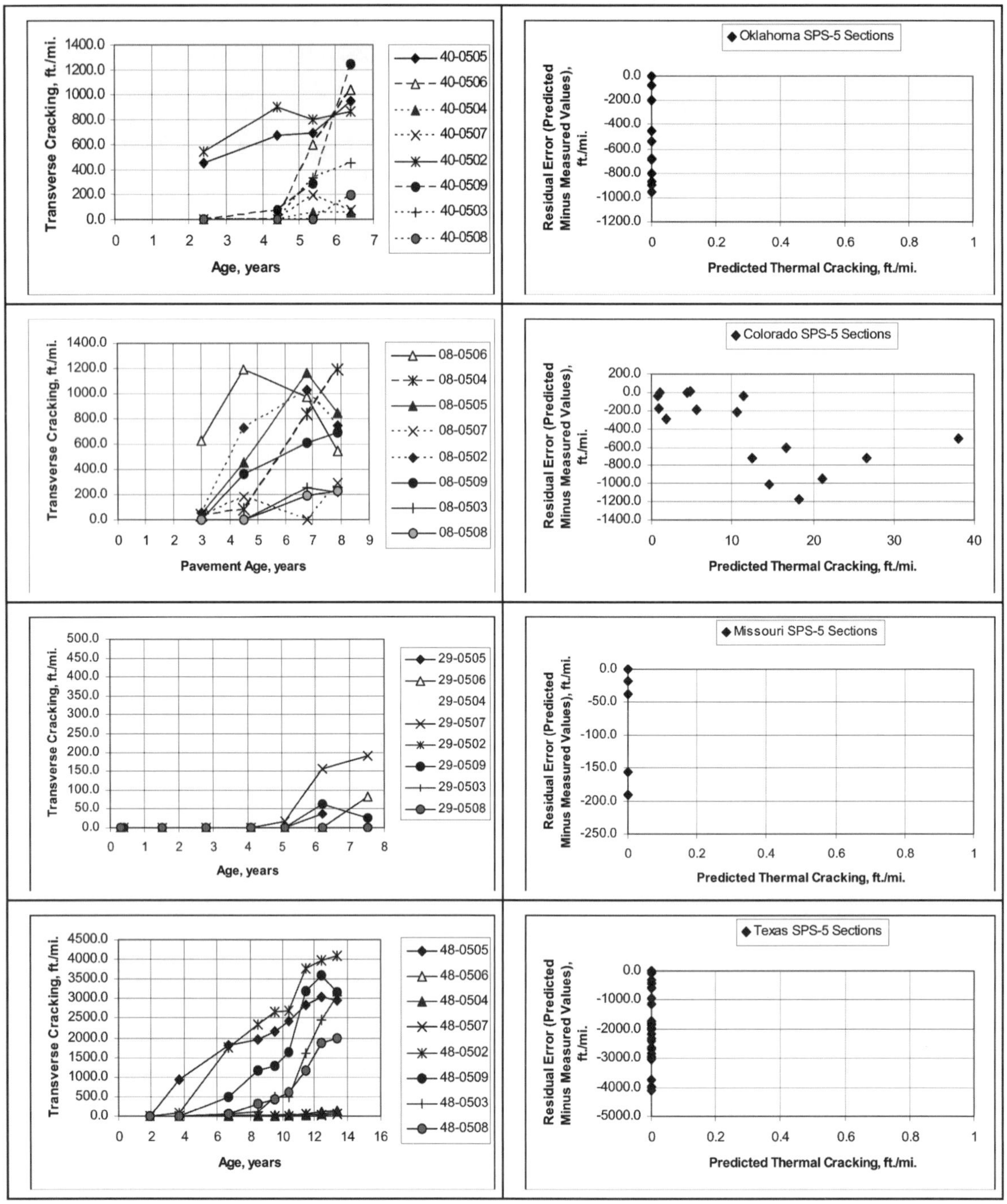

Figure A2.4.D.2-3. Transverse Cracking Measurements and Thermal Cracking Predictions with the Global Calibration Values

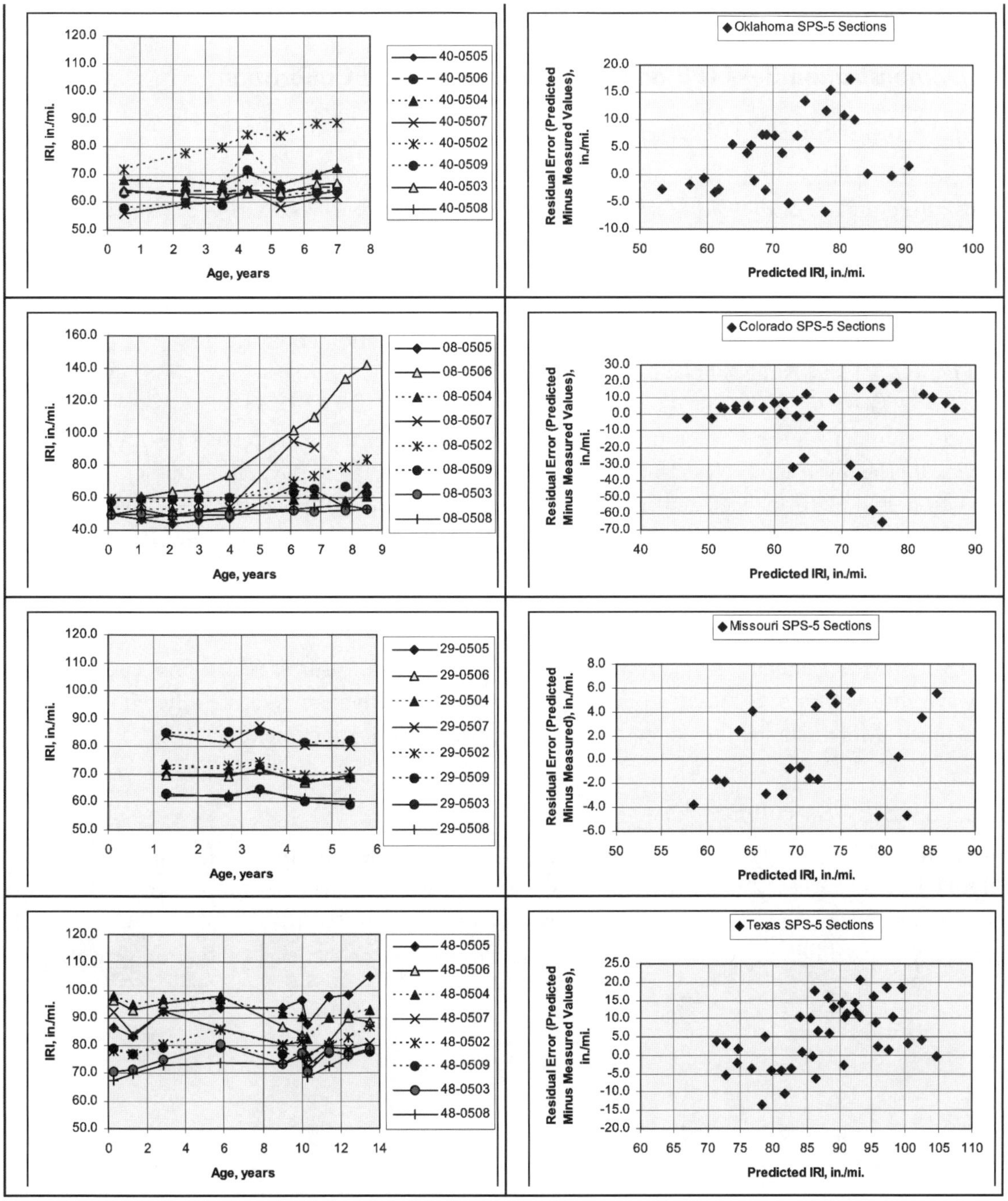

Figure A2.4.D.2-4. IRI Measurements and Predictions with the Global Calibration Values

A3. New Rigid Pavements—Jointed Plain Concrete Pavements

A3.1 Demonstration 3—LTPP and PMS Data and Local Calibration

A3.1.1 Description of LTPP And PMS Segments

The Missouri Department of Transportation (MODOT) identified and selected 24 projects for demonstrating the use of the *Local Calibration Guide of the Mechanistic-Empirical Pavement Design Guide* (MEPDG). Six of the projects were from the Long-Term Pavement Performance (LTPP) program and 18 were MODOT pavement management (PM) segments. All of the projects were newly constructed or reconstructed jointed plain concrete pavements (JPCP) with construction date ranging from 1955 to 1998. Table A3-1 provides general information about the projects selected, while Figure A3-1 shows the general location of each project.

Table A3-2 summarizes the selected projects' structures (i.e., layer types, material types, and layer thicknesses). The information assembled was extracted from LTPP and MODOT traffic, materials, construction, and performance databases and other construction records. With the exception of LTPP 0701 which was constructed in 1955, all of the JPCPs selected were doweled. Dowel diameter ranged from 1.25 to 1.5 in. Joint spacing ranged from 15 to 30 ft, while slab widths ranged from 10 to 14 ft. The JPCP projects had both tied and non-tied Portland cement concrete (PCC) shoulders.

Table A3-3 provides general information about the JPCPs' design features. For all projects, a surface short-wave absorptivity of 0.85 and a permanent curl/warp effective temperature difference of –10°F were assumed. More detailed project descriptions are provided in Attachment A3.2.A.

A3.1.2 Step-by-Step Procedure

The steps included in the *Local Calibration Guide* were followed for this demonstration using MODOT PM-roadway segments and LTPP test sections. Many of the decisions made by MODOT were based on an expedited time frame to collect the necessary data for the demonstration. MODOT would likely make different decisions given a longer time frame for the local calibration process.

Table A3-1. General Information for the Selected Missouri LTPP and PMS Projects

ARA ID	Pave-ment Type	Project ID (MODOT PM ID)	Route	Direction	County	Number and Length of Homogenous Sections	Initial AADTT (Design Lane)	Begin MP	End MP	Long.	Lat.	Analysis Period (yr)
0701	JPCP	LTPP SPS-7	SR-6	East	Buchanan	1	244	—	—	—	—	—
0807	JPCP	LTPP SPS-8	SR-6	West	Buchanan	1	80	—	—	37.91	–90.66	1998–2008
0808	JPCP	LTPP SPS-8	US-54	West	Callaway	1	80	—	—	37.91	–90.66	1998–2008
A601	JPCP	LTPP SPS-6	SR-TT	East	Boone	1	117	—	—	36.96	–93.22	1968–2008
A807	JPCP	LTPP SPS-8	IS-35	South	Harrison	1	100	—	—	37.91	–90.66	1998–2008
A808	JPCP	LTPP SPS-8	US-71	South	Newton	1	180	—	—	37.91	–90.66	1998–2008
A02-E	JPCP	BRS-44(11)	US-71	South	Newton	1	10	4.14	4.24	39.78	–94.73	1987–2008
A02-W	JPCP	BRS-44(11)	US-36	West	Caldwell	1	10	204.63	204.73	39.78	–94.73	1987–2008
A03-W	JPCP	5P5434	US-24	East	Marion	3	453	85.27	85.37	38.78	–92.02	1975–2008
D03-E	JPCP	J5P0381	US-54	West	Callaway	1	166	2.00	2.10	38.96	–92.39	1995–2008
E01-S	JPCP	J1I0541	US-54	East	Callaway	3	1120	17.57	17.67	40.32	–94.01	1996–2008
E03-S	JPCP	J7P0489	US-54	East	Callaway	2		281.12	281.22	36.95	–94.43	1997–2008
E04-S	JPCP	J7P0490	SR-7	South	Henry	2	665	286.91	287.01	36.87	–94.43	1997–2008
F01-W	JPCP	J1P0489B	US-60	East	Butler	4	592	154.01	154.11	39.74	–94.03	1996–2008
F02-E	JPCP	J3P0284	US-60	West	Butler	1	462	212.80	212.90	39.93	–91.48	1997–2008
F04-W	JPCP	J5P0410	SR-74	East	Cape Girardeau	1	484	83.66	83.76	38.81	–91.99	1994–2008
F05-E	JPCP	J5P0411C	US-63	South	Callaway	3	429	183.74	183.84	38.76	–92.05	1993–2008
F06-E	JPCP	J5P0412C	SR-M	East	Jefferson	3	520	177.32	177.42	38.73	–92.06	1994–2008
F07-S	JPCP	J4P0861D	SR-6	East	Buchanan	2	779	72.23	72.33	38.41	–93.85	1997–2008
F08-E	JPCP	J0P0571	SR-6	West	Buchanan	2	396	269.62	269.62	36.79	–90.36	1997–2008
F09-W	JPCP	J0P0572	US-54	West	Callaway	2	775	64.99	65.09	36.79	–90.25	1997–2008
F10-E	JPCP	J0U0412C	SR-TT	East	Boone	1	278	7.82	7.92	37.29	–89.56	1995–2008
G01-S	JPCP	J5P0621	IS-35	South	Harrison	2	123	144.80	144.90	38.63	–92.20	1994–2008
G03-E	JPCP	J6S064I	US-71	South	Newton	3	200	3.41	3.51	38.36	–90.45	1998–2008

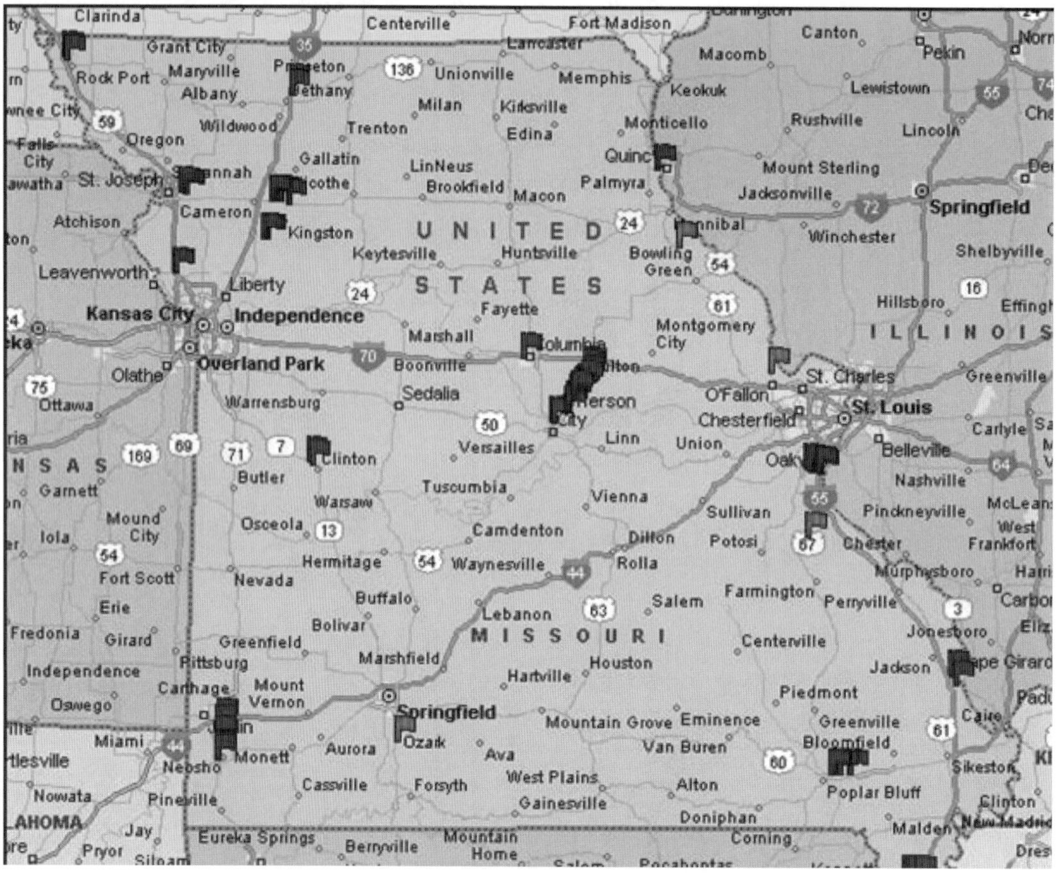

Figure A3-1. General Location of the Missouri LTPP and PMS Roadway Segments Selected for Demonstrating the Local Validation-Calibration Process

Step 1—Select Hierarchical Input Level

The appropriate MEPDG hierarchical input Level (1 through 3) used in the local validation-calibration process was selected to be consistent with MODOT's procedures for determining MEPDG inputs for day-to-day use. This demonstration utilizing LTPP and PMS-roadway segments used Input Levels 2 and 3 data that were available for typical designs. Level 1 inputs were generally unavailable for most of these projects, because of their age. The general information from which the inputs were determined for each input category is discussed in Step 5.

Table A3-2. Material Types and Layer Thicknesses for the Missouri LTPP and PMS Segments

Project ID	Layer No.	Layer Type	Material Type	Thickness, in.
LTPP 0701	1	PCC	JPCP	8
	2	Granular	A-2-4	4
	3	Subgrade	A-7-6	Semi-Infinite
LTPP 0807	1	PCC	JPCP	7.9
	2	Granular	A-2-4	6.3
	3	Subgrade	A-7-6	Semi-Infinite

Table A3-2—Continued

Project ID	Layer No.	Layer Type	Material Type	Thickness, in.
LTPP 0808	1	PCC	JPCP	10
	2	Granular	A-2-4	6.3
	3	Subgrade	A-7-6	Semi-Infinite
LTPP A807	1	PCC	JPCP	8.6
	2	Granular	A-2-4	6
	3	Subgrade	A-7-6	Semi-Infinite
LTPP A808	1	PCC	JPCP	11.1
	2	Granular	A-2-4	6
	3	Subgrade	A-7-6	Semi-Infinite
A02-E (all sampling units)	1	PCC	JPCP	8
	2	Granular	A-2-4	4
	3	Subgrade	A-6	Semi-Infinite
A02-W (all sampling units)	1	PCC	JPCP	8
	2	Granular	A-2-4	4
	3	Subgrade	A-6	Semi-Infinite
D03-E (all sampling units)	1	PCC	JPCP	8
	2	Granular	A-2-4	4
	3	Subgrade	A-6	Semi-Infinite
E01-S (all sampling units)	1	PCC	JPCP	8
	2	Stabilized	Cement Stabilized	4
	3	Subgrade	A-2-4	4
	4	Subgrade	A-6	Semi-Infinite
E03-S (all sampling units)	1	PCC	JPCP	8
	2	Stabilized	Cement Stabilized	4
	3	Subgrade	A-2-4	4
	4	Subgrade	A-6	Semi-Infinite
E04-S (all sampling units)	1	PCC	JPCP	8
	2	Stabilized	Cement Stabilized	4
	3	Subgrade	A-2-4	4
	4	Subgrade	A-6	Semi-Infinite
F01-W (all sampling units)	1	PCC	JPCP	8
	2	Granular	A-2-4	4
	3	Subgrade	A-6	Semi-Infinite
F02-E (all sampling units)	1	PCC	JPCP	8
	2	Granular	A-2-4	4
	3	Subgrade	A-6	Semi-Infinite
F04-W (all sampling units)	1	PCC	JPCP	8
	2	Granular	A-2-4	4
	3	Subgrade	A-6	Semi-Infinite

Table A3-2—Continued

Project ID	Layer No.	Layer Type	Material Type	Thickness, in.
F05-E (all sampling units)	1	PCC	JPCP	8
	2	Granular	A-2-4	4
	3	Subgrade	A-6	Semi-Infinite
F06-E (all sampling units)	1	PCC	JPCP	8
	2	Granular	A-2-4	4
	3	Subgrade	A-6	Semi-Infinite
F07-S (all sampling units)	1	PCC	JPCP	8
	2	Granular	A-2-4	4
	3	Subgrade	A-6	Semi-Infinite
F08-E (all sampling units)	1	PCC	JPCP	8
	2	Granular	A-2-4	4
	3	Subgrade	A-6	Semi-Infinite
F09-W (all sampling units)	1	PCC	JPCP	8
	2	Granular	A-2-4	4
	3	Subgrade	A-6	Semi-Infinite
F10-E (all sampling units)	1	PCC	JPCP	8
	2	Granular	A-2-4	4
	3	Subgrade	A-6	Semi-Infinite
G01-S (all sampling units)	1	PCC	JPCP	8
	2	Granular	A-2-4	4
	3	Subgrade	A-6	Semi-Infinite
G03-E (all sampling units)	1	PCC	JPCP	8
	2	Granular	A-2-4	4
	3	Subgrade	A-6	Semi-Infinite

Step 2—Experimental Factorial and Matrix or Sampling Template

A detailed sampling template was created for this demonstration, which is shown in Table A3-4 along with the number of LTPP and PMS segments or projects within each cell (refer to Step 3). This sampling template identified the different features that have been used by MODOT over time. Some of these features, however, are no longer commonly used in current practice.

As extreme climate variations and soil conditions do not occur across Missouri, with the exception of some localized areas, the primary tier in the factorial was the JPCP design features (e.g., slab width, edge support, and etc.). The number of roadway segments selected for the sampling template should result in a balanced factorial with the same number of replicates within each cell. As shown in Table A3-4, the factorial is highly unbalanced and most cells do not have replication. Thus, it was decided to validate the global calibration values for the common design strategy and features used by MODOT (the cell in Table A3-4 with the majority of roadway segments).

Table A3-3. Design Features for the Missouri LTPP and PMS Segments

Project Name	Joint Spacing, ft	Dowel Diameter, in.	Tied PCC Shoulder (Y or N?)	Widened Slab (Y or N?)	Slab Width, ft
0701	20	0	N	N	12
0807	15	1.25	N	N	10
0808	15	1.25	N	N	10
A601				N	12
A807	30	1.25	N	N	10
A808	30	1.25	N	N	10
A02-E	30	1.25	N	N	12
A02-W	30	1.25	N	N	12
D03-E	15	1.25	Y	N	14
E01-S	15	1.5	Y	N	14
E03-S	15	1.5	Y	N	14
E04-S	15	1.5	Y	N	14
F01-W	15	1.5	Y	N	14
F02-E	15	1.5	Y	N	14
F04-W	15	1.5	Y	N	14
F05-E	15	1.5	Y	N	14
F06-E	15	1.5	Y	N	14
F07-S	15	1.5	Y	N	14
F08-E	15	1.5	Y	N	14
F09-W	15	1.5	Y	N	14
F10-E	15	1.5	Y	N	14
G01-S	15	1.5	Y	N	14
G03-E	15	1.5	Y	N	14

Table A3-4. Simplified Sampling Template for the Demonstration Using LTPP and PM Data

Edge Support Conditon	Slab Width, ft	Joint Spacing, ft	Dowel Diameter, in.					
			None		1.25		1.5	
			PCC Thickness, in.					
			>10	>10	<10	>10	<10	>10
None	10	15			2			
		30			1	1		
	12	20	1					
		30			2			
Tied Shoulder	10	15						
		30						
	12	15						
		30						
	14	15			1		16	
		30						

Step 3—Estimate Sample Size for Each Performance Indicator Prediction Model
A 90-percent level of significance was used for estimating the sample size (total number of roadway segments or projects). Higher confidence levels can be used, but that will increase the number of segments needed. Table A3-5 summarizes the estimated sample size for each measure of performance. The tolerable bias was estimated from the selected significance level (90 percent) for estimating the sample size and the MEPDG distress/IRI prediction model standard error of estimate (SEE).

Table A3-5. Estimated Number of PMS Segments Needed for the Local Validation-Calibration Process

Performance Indicator	Performance Indicator Threshold (at 90% Reliability)	Tolerable Bias	Standard Error of Estimate (SEE)	Minimum Number of Projects Required for Calibration-Validation
Transverse cracking, percent slabs cracked	6.1 (15)	7.2	4.5%	11
Transverse joint faulting, in.	0.078 (0.12)	0.1	0.033, in.	13
Smoothness, in./mi	110 (160)	27.5	17.1, in./mi	87

Note: $n = \left(\frac{Z_{\alpha/2}\sigma}{E}\right)^2$

where:
$Z_{\alpha/2}$= 1.601 (for a 90 percent confidence interval),
s = design criteria, and
E = bias at 90-percent reliability (1.601*SEE).

Step 4—Select Roadway Segments
Projects should be selected to cover a reasonable range of distress and smoothness values. Roadway segments exhibiting premature or accelerated distress levels, as well as those exhibiting superior performance (low levels of distress over long periods of time), can be used, but with caution. The roadway segments selected for the sampling template when using hierarchal input Level 3 should represent average performance conditions.

A limited number of potential LTPP and PMS segments were available for this demonstration, as many of the MODOT PMS segments had insufficient data to be included in the local validation-calibration effort. In addition, PMS roadway segments with the other features identified in Table A3-4 are simply unavailable. Thus, it was decided to demonstrate the validation-calibration of the MEPDG for the standard JPCP design being used by MODOT. It is expected that many agencies will have similar policies and data limitations.

Twenty-four segments with multiple distress measurements within each segment were selected for this example (see Figure A3-1). As shown in Table A3-5, the 24 segments are sufficient for performing local validation-calibration of the transverse cracking and transverse joint faulting models. Although more segments are required for local calibration of the IRI model, the 24 segments (16 of those located in one cell), were used for this demonstration. Although it is important that each roadway segment have repeated measures of distress over several years, it would not be good practice to have some segments with several repeated measures (about 10 observations over 10 years), while other segments have very few measurements (about two or three observations over 10 years). This is because the segments with more repeated measures will have better estimates of pavement distress and IRI while those with few measurements will be less accurate and thus introduce greater error into the validation-calibration process.

Step 5—Extract and Evaluate Distress and Project Data

This step is grouped into four activities: (1) extracting and reviewing the distress/IRI data; (2) comparing the performance indicator magnitudes to the trigger values; (3) evaluating the distress data to identify anomalies and outliers; and (4) determining the inputs to the MEPDG.

Step 5.1—Extract, Review, and Convert Measured Values to the Values Predicted by the MEPDG, if Needed.

First, the distress or performance indicator measurements included in the MODOT PMS databases were reviewed to determine whether the measured values are consistent with the values predicted by the MEPDG. For both sources of data, the measured transverse cracking values are different, while the transverse joint faulting and IRI values are similar and assumed to be the same. The transverse cracking values and how they were used in the local calibration process are defined below.

- **Transverse Cracking.** The MEPDG requires the percentage of all slabs with mid-panel fatigue transverse cracking. Both MODOT and LTPP describe transverse cracking as cracks that are predominantly perpendicular to the pavement slab centerline. Measured cracking is reported in three severity levels (low, medium, and high) and provides distress maps showing the exact location of all transverse cracking identified during visual distress surveys. Thus, the databases contains, for a given number of slabs within a 500-ft pavement segment, the total number of low-, medium-, and high-severity transverse cracks. Since LTPP does not provide details on whether a given slab has multiple cracks, as shown in Figure A3-2, a simple computation of percent slabs with this kind of data can be misleading. Therefore, in order to produce an accurate estimate of percent slabs cracked, distress maps or videos prepared as part of distress data collection were reviewed to determine the actual number of slabs with transverse "fatigue" cracking for the 500-ft pavement segments. Total number of slabs was also counted. Percent slabs cracked was defined as follows:

$$\text{Percent Slabs Cracked} = \left(\frac{\text{Number of cracked slabs}}{\text{Total number of slabs}}\right) \times 100 \qquad (A3\text{-}1)$$

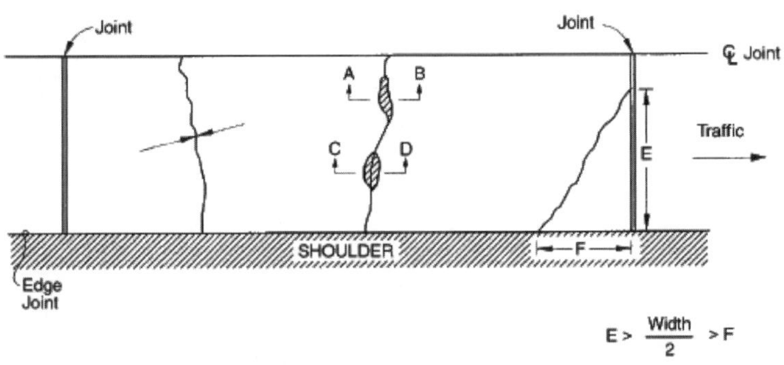

Figure A3-2. LTPP Transverse Cracking

- **Transverse Joint Faulting**. Transverse joing cracking is measured and reported by MODOT and LTPP as the difference in elevation to the nearest 1 mm between the pavement surfaces on either side of a transverse joint. The mean joint faulting for all joints within a 500-ft pavement section is reported. This is comparable to the MEPDG predicted faulting.
- **IRI**. The values included in the MODOT PMS database are comparable to the MEPDG predicted IRI.

The average measured distress/smoothness value should be determined for each measurement period for each LTPP and PMS segment. The time-history data should not be smoothed but can be cleaned to remove data errors. As an example, measured distresses that are recorded as zero after multiple values of distress have been recorded. Plots of the average time-history data are included in Attachment A3.2.B. Some important observations of the data that have an impact on the validation-calibration process are listed below.

- Some measurement errors are present for all performance indicators. The measured values sometime increase and decrease with time. In addition, there are abrupt changes in the faulting and IRI data over time. Thus, improving on the precision of the prediction model is not likely.
- Few of the PMS segments have any measured fatigue cracking. Thus, confirming the transverse cracking global calibration values using statistical methods is not likely.
- Faulting is low for most of the LTPP and PMS segments included within this demonstration. Thus, confirming the faulting global calibration values will be limited to faulting significantly less than the design criteria or trigger value.

Step 5.2—Compare Distress Magnitudes to Trigger Values
The next activity of this step is to compare the distress magnitudes to the trigger values for each distress. Then answer the question—does the sampling matrix include values close to the design criteria or trigger value? Table A3-6 summarizes the average, maximum, and minimum distress values for each performance indicator as compared to the distress levels that trigger a major rehabilitation activity.

Table A3-6. Summary of the Maximum Values of Different Performance Indicators in Comparison to the Design Criteria or Trigger Values (Number of Sites = 24)

Distress or Performance Indicator	Design Criteria	Maximum Values Statistics			
		Average Max. Value	Lowest Max. Value	Largest Max. Value	Standard Deviation of Max. Values
Transverse cracking, percent slabs cracked	Mean: 6.1 90% Rel: 15	6	0	78	17.1
Transverse joint faulting, in.	Mean: 0.078 90% Rel: 0.12	0.014	0	0.13	0.018
Smoothness, in./mi	Mean: 110 90% Rel: 160	99.5	45.8	204.6	35.5

As shown in Table A3-6, mean cracking and IRI values are close to the design criteria, while the mean faulting value was significantly less than the mean design criteria. In fact, the average faulting distress magnitudes are more than two standard deviations below the design criteria. This comparison suggests that the values used as MODOT's faulting distress design trigger values are too high or that JPCPs in Missouri are generally rehabilitated for other reasons. This observation becomes important when evaluating the bias and standard error terms of the prediction models under Steps 7 and 9, respectively. Note that the mean value of faulting may be too low to validate and accurately determine the local calibration values or adjustments for predicting the increase in faulting over time; especially when 0.078-in. (0.12-in.) faulting was selected for the design criteria.

Step 5.3—Evaluate Distress Data to Identify Anomalies and/or Outliers
The distress data for each roadway segment included in the sampling template should be looked at prior to determining all of the MEPDG input parameters. This evaluation can be limited to visual inspection of the data over time to ensure that the distress data are reasonable (i.e., time-history plots), or include a detailed statistical comparison of the performance data. As a minimum, the following questions should be asked (Attachment A3.2.B includes graphs that show the distress values over time for the roadway segments).

- Does the data make sense within and between each roadway segment? All of the data extracted from the Missouri LTPP and PMS segments looked reasonable and appeared to represent typical performance characteristics and conditions. Obviously, any zeros that represent non-entry values should be removed from the local validation-calibration database. Distress data that return to zero values within the measurement period may indicate some type of maintenance or rehabilitation activity.
  - Measurements taken after structural rehabilitation should be removed from the database or the observation period should end prior to the rehabilitation activity.
  - Distress values that are zero as a result of some maintenance or pavement preservation activity, which is a part of the agency's management policy, should be removed.
- Are there segments with anomalies, outliers, or blunders in the data? If the outliers or anomalies can be explained and are a result of some non-typical condition, they should be removed. If the outlier or anomaly cannot be explained, they should remain in the database. No outliers or anomalies were found in the LTPP and MODOT data (see Attachment A3.2.B).

Step 5.4 —Inputs to the MEPDG for Each Input Category
The following provides a brief discussion of the information extracted from the LTPP and MODOT databases and files for determining the inputs needed to execute the MEPDG for each of the PMS segments.

- General inputs:
  - Initial average annual daily truck traffic (AADTT) and traffic growth rate were computed using historical truck traffic counts obtained from LTPP and MODOT.
  - Initial IRI was backcasted using historical IRI data provided by MODOT (see Figure A3-3 for an illustration).
  - **Construction Histories, Cross Sections, and Layer Thicknesses**—As-built plans that were available from MODOT and LTPP records were used to determine the material types

and layer thicknesses for each LTPP and PMS segment. It was assumed that all layers were placed in accordance with the project specifications. Material properties consistent with the specifications were used for inputs with minimum construction data.

- The construction date was obtained from MODOT as-built plans or the construction database files or from LTPP database.
- The opening month to traffic for all projects was assumed to be one month following the PCC placement.
- **Traffic**—Site specific values were used for all input (number of lanes, traffic growth, truck traffic classification groups or vehicle class distribution, initial AADTT, and axle-load distribution). Default values were adopted for other traffic inputs such as axle configuration and speed.
- **Climate**—The longitude and latitude of each LTPP or MODOT PMS segment was used to create a virtual weather station for that segment of roadway. The weather stations in Missouri and adjacent states were used to create the virtual weather stations.

- Non-LTPP JPCP:
  - Each sample unit was cored and the cores tested in the lab for the following:
    - PCC layer thicknesses.
    - PCC compressive strength and coefficient of thermal expansion.
    - PCC elastic modulus and flexural strength was obtained through the use of state-specific correlations based on PCC compressive strength. The correlations are presented as follows:

$$E_{PCC} = 64200 * \sqrt{f_c^{'}} \quad \text{(A3-2)}$$

$$MR = 9.6 * \sqrt{f_c^{'}} \quad \text{(A3-3)}$$

  - State-specific defaults of unbound material properties (Atterberg limits, gradation, and so on) obtained from MODOT records were used for material characterization.
  - Initial AADTT and traffic growth rate was computed using historical truck traffic counts obtained from MODOT.
  - Initial IRI was backcalculated using historical IRI data provided by MODOT.
  - FWD deflection test data were provided by MODOT. Using these data, pavement layer moduli and modulus of subgrade reaction were backcalculated. Rigid pavement backcalculation was done based on the AASHTO AREA method.
  - Automated and manual distress surveys were conducted by MODOT and the information provided to ARA. The data were used to characterize pavement condition (transverse cracking, faulting, and IRI).

- LTPP JPCP:
  - Review of LTPP traffic, inventory, materials, and distress databases.
  - Key inputs were obtained as described below:
    - PCC strength and modulus: Depending on the project type (SPS-8 versus GPS sections), the 14-, 28-, 365-day or long-term modulus of rupture (flexural strength), elastic modulus, compressive strength, and so on were tested and are available in the LTPP database. The available data were used to estimate Level 1 MEPDG inputs (i.e., 14-, 28-, and 90-day MR and E_{PCC}) for SPS projects and Level 3 MEPDG inputs (i.e., 28-day MR and E_{PCC}) for GPS projects.
    - PCC coefficient of thermal expansion (CTE): For both the SPS and GPS projects PCC materials, CTE values were measured by LTPP.
    - Unbound aggregate materials and soils inputs for climate modeling: These were determined using the LTPP lab tested gradation and Atterberg limit values.
    - Resilient modulus of unbound aggregate materials used as base or subbases: Default MEPDG values were adopted based on the material AASHTO soil classification determined using LTPP lab tested gradation and Atterberg limit values.
    - Subgrade resilient modulus (tested at optimum moisture): Obtained through backcalculation using FWD deflection test data provided by LTPP. Rigid pavement backcalculation was done based on the AASHTO AREA method. For flexible pavements, backcalculation was done using EVERCALC.

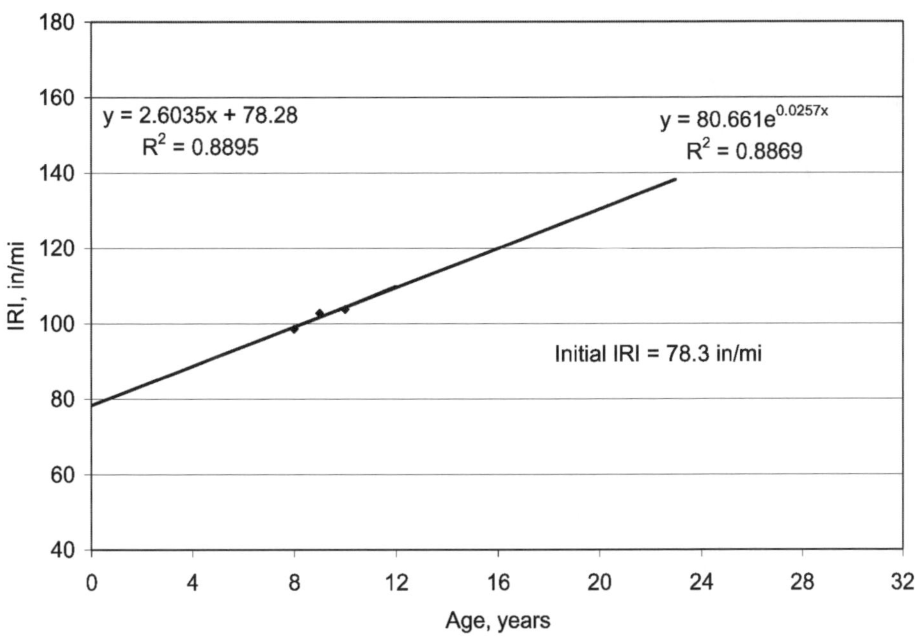

Figure A3-3. Illustration of Initial IRI Backcasting for G01-S-S1

For both non-LTPP and LTPP-JPCP, subgrade lab resilient modulus (M_r) at optimum moisture content is the required input when the Integrated Climatic Model (ICM) is used to determine the seasonal effects over time. This input was used in the program to backcalculate a *k*-value for each month for rigid and composite pavements. For flexible pavements, M_r is seasonally adjusted and the seasonally adjusted values used directly as inputs for analysis.

FWD data from the LTPP database or MODOT were used to backcalculate a long-term subgrade *k*-value as appropriate. The point in time chosen for the backcalculation was selected to represent a long-term value (presumably when equilibrium moisture contents are reached in the field) when the subgrade is not saturated or frozen (summer months). An appropriate subgrade lab resilient modulus (M_r) at optimum moisture content value was then selected through trial and error to obtain an MEPDG estimate of the long-term in situ *k*-value which was similar to the field tested value.

Step 6—Conduct Field and Forensic Investigations
If the assumptions in the MEPDG are challenged by the agency, forensic investigations may be needed to measure and quantify input parameters such as permanent curl/warp effective temperature. An example of forensic field measurements to determine permanent curl/warp effective temperature for an in-service JPCP is shown in Figure A3-4.

Figure A3-4. Illustration of Field Measurement of Permanent Curl/Warp Effective Temperature

Another reason for forensic examination is reasonableness of input data. For example, if measured distress, design, materials, and other data is found to be unreasonable, a forensic investigation may be warranted.

LTPP 0701 reports approximately 0.125 in. of faulting after 55 years in service. This seems to be quite low for a highly trafficked undoweled JPCP. Thus, it is hypothesized that the faulting for the only undoweled JPCP used in local calibration is unreasonable. However, a detailed review of MODOT construction, maintenance, and rehabilitation records indicated no such repairs over the 55-year history of the pavement. A detailed forensic investigation of the segment is thus warranted to determine the accuracy of the measured faulting. For this demonstration, it was assumed that the faulting data gathered was accurate and was thus included in the local calibration demonstration.

Steps 7 through 11 should be executed for each specific MEPDG predicted distress after the forensic investigations as needed are completed. For this demonstration, MODOT decided to accept the assumptions and conditions included in the MEPDG for the global calibration effort (i.e., permanent curl/warp effective temperature = –10). Thus, no field and forensic investigations were planned or conducted to determine parameters such as permanent curl/warp effective temperature.

Step 7—Assess Local Bias from Global Calibration Factors
The MEPDG was executed using the global calibration values to predict the performance indicators for each LTPP and PMS segment. The predicted performance measures are shown in Attachment A3.2.B relative to the measured values for the LTPP and PMS segments. Table A3-7 lists the bias for each performance indicator and whether the hypothesis is rejected or accepted, while Figures A3-5 and A3-6 compare the predicted and measured values for each performance indicator.

Table A3-7. Summary of the Statistical Parameters—Global Calibration Values Used for Predicting Performance Indicators for the Missouri LTPP and PMS Sections

Performance Indicator	Project	Bias (p-value)	Standard Error	R^2 (See Note 1)	Hypothesis; H_o: $y_i - \hat{y}_i = 0$	Comment
Transverse cracking	New	0.5283	—	Poor	Accept	No bias
Transverse joint faulting	New	< 0.0001	0.03 in.	0.71	Reject	Bias. The hypothesis should be re-evaluated after recalibration to reduce bias
IRI	New	0.1478	11.1 in./mi	0.85	Accept	No bias

Notes: 1. Residual Error = $y_i - \hat{y}_i$ where y_i Measured or Observed Value and $\hat{y}_i$ = Predicted Value.
2. Poor means that the model did not explain variation in the measured data within and between the LTPP and PMS segments.
3. The null hypothesis is that the average residual error or bias is zero for a specified confidence level or level of significance. 90%-confidence level was used in this demonstration. In other words: $H_O : \sum\left(y_i - \hat{y}_i\right) = 0$

The information presented in Table A3-7 shows that the MEPDG global calibration values resulted in a significant bias for transverse joint faulting. The bias for transverse cracking and IRI was not significant. R^2 for transverse cracking could not be estimated since measured cracking was mostly zero.

The following summarizes the findings for each performance indicator.

Transverse Cracking Predictions

- There are too few LTPP and PMS segments with appreciable transverse fatigue cracking to validate or confirm use of the global calibration values and determine local adjustment or calibration values using statistical methods. The model was, therefore, evaluated by categorizing measured and predicted slab transverse cracking into eight groups and determining how often the measured and predicted cracking fell within the same group. Group characteristics were determined based on the range of transverse cracking measured from all projects evaluated. The results of the comparisons are summarized in Table A3-8.

  As shown in Table A3-8, a vast majority of the measured and predicted transverse cracking (approximately 74 percent) fell within the same measured and predicted transverse cracking grouping. All of these were for pavements with very little cracking distress. For the moderately to highly distressed pavements (measured cracking ranged from 60 to 80 percent slabs), prediction ranged from 20 to 100 percent. This indicates that cracking was predicted less accurately for moderately to highly distressed pavements. The levels of cracking predicted are believed to be reasonable as model accuracy is more important for lower levels of the distress close to the design criteria. The JPCP transverse cracking was therefore found to predict the distress reasonably well for MODOT pavements.

Table A3-8. Comparison of Measured and Predicted Transverse Slab Cracking (Percentage of All Data Points)

Measured Percent Slabs Cracked	MEPDG Predicted Percent Slabs Cracked							
	0–2	2–5	5–10	10–20	20–40	40–60	60–80	80–100
0–2	74	0	0	0	0	0	0	0
2–5	4	0	0	0	0	0	0	0
5–10	0	0	0	0	0	0	0	0
10–20	1	0	0	0	0	0	0	0
20–40	0	0	0	0	0	0	0	0
40–60	0	0	0	0	0	0	0	3
60–80	5	0	0	0	3	8	0	3
80–100	0	0	0	0	0	0	0	0

Note: Total data points = 76.

Transverse Joint Faulting Predictions

- Extensive error between the predicted and measured faulting was observed. The error observed was in the form of the MEPDG model over predicting faulting from some pavement segments. No case of significant under prediction of faulting was observed. A good correlation exists between the predicted and measured faulting values (refer to Figure A3-5).

- A detailed evaluation of predicted and measured faulting showed that doweled JPCP with joint spacings less than 20 ft exhibited very little faulting (faulting ranged from 0 to 0.04 in.). The range of measured faulting of 0 to 0.04 in. is typical for JPCP with relatively short joint spacings and JPCP with properly doweled joints. The review also showed that the MEPDG over-predicts faulting for JPCP that are not doweled or have long joint spacings (i.e., joint spacing >20 ft).

IRI Predictions

- Although the IRI predictions are heavily dependent on the other distresses calculated by the MEPDG such as faulting and the site factor term, the model was still evaluated to determine robustness and reasonableness. Figure A3-6 shows a plot of predicted and measured JPCP IRI. The information presented shows that the global coefficients for the JPCP IRI model predict IRI adequately despite the bias observed in MEPDG predicted faulting. This is because the bias in predicted faulting was limited to a few segments and has limited impact on overall IRI predictions.

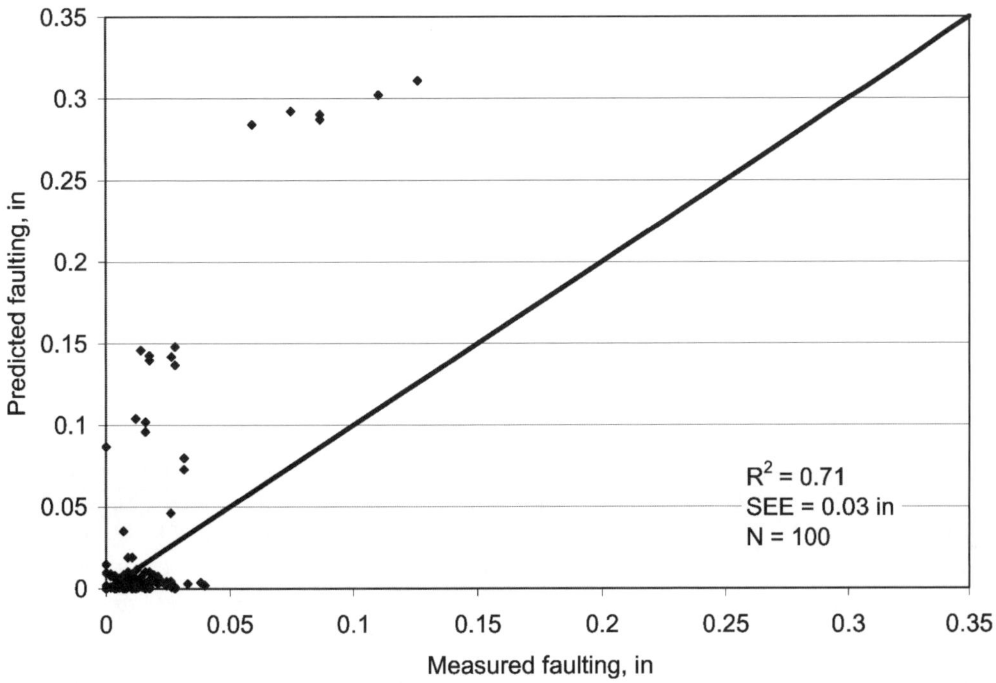

Figure A3-5. Comparison of Predicted and Measured Transverse Joint Faulting Using the Global Calibration Values

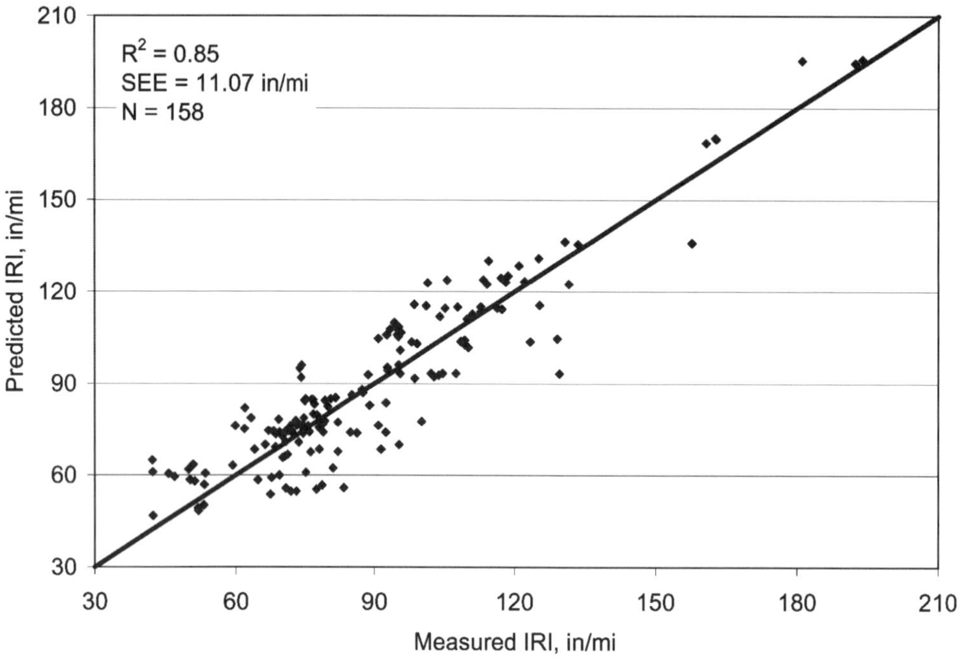

Figure A3-6. Comparison of Predicted and Measured IRI Using the Global Calibration Values

Step 8—Eliminate Local Bias of Distress Prediction Models

The process used to eliminate the bias found to be significant from using the global calibration values depends on the cause of the bias and the accuracy desired by the agency. In general, there are three possibilities which are listed below.

1. The residual errors are, for the most part, always positive or negative with a low standard error of the estimate in comparison to the trigger value, and the slope of the residual errors versus predicted values is relatively constant and close to zero. The precision of the prediction model is reasonable but the accuracy is poor. In this case, the local calibration coefficient is used to reduce the bias. This condition generally requires the least level of effort and the fewest number of runs or iterations of the MEPDG to reduce the bias.

2. The bias is low and relatively constant with time or number of loading cycles, but the residual errors have a wide dispersion varying from positive to negative values. The accuracy of the prediction model is reasonable, but the precision is poor. In this case, the coefficient of the prediction equation is used to reduce the bias but the value of the local calibration coefficient is probably dependent on some site feature, material property, and/or design feature included in the sampling template. This condition generally requires more runs and a higher level of effort to reduce the bias.

3. The residual errors versus the predicted values exhibit a significant and variable slope that appears to be dependent on the predicted value. The precision of the prediction model is poor and the accuracy is time or number of loading cycles dependent—there is poor correlation between the predicted and measured values. This condition is the most difficult to evaluate because the exponent of the number of loading cycles needs to be considered. This condition also requires the highest level of effort and many more runs to reduce the bias.

The agency should first decide on whether to use the agency specific values or the local calibration factors that are considered as inputs in the MEPDG software. Either one can be used with success—for this demonstration the local calibration parameters were used. Time-history plots of each performance indicator should be prepared to determine if one or multiple calibration factors need to be evaluated, as noted above. The following describes the process used to eliminate the bias in predicted transverse joint faulting.

Transverse Joint Faulting Predictions

A good correlation was found between the predicted and measured faulting using the global calibration values (refer to Table A3-7 and Figure A3-7). The causes of bias in predicted transverse joint faulting were identified in completing an analysis of the residual errors relative to the sampling template.

- Faulting was under-predicted for segments with no dowel or with joint spacings greater than 20 ft.
- There was the need to modify the faulting model global coefficients (recalibrate) to reduce the observed bias.
- A review of the effect of the faulting model global coefficients on predicted faulting showed the following:
    - The calibration coefficients can be divided into two groups, namely:
        - Coefficients affecting the shape of the faulting vs. traffic or faulting vs. time prediction curve (C_1, C_2, C_3, C_4, and C_7), which can be further subdivided into three groups as follows:
            - Coefficients affecting the magnitudes of mid-range faulting (C_1 and C_2).
            - Coefficients affecting the rate of the initial faulting development (C_3 and C_4).
            - Coefficient impacting the slope of faulting in the long term (C_7).
        - Coefficients magnifying the effects of design features and site conditions (C_6, C_7, and C_8).

Figure A3-7 illustrates different parts of the faulting development. Figures A3-8, A3-9, A3-10, and A3-11 present sensitivity of predicted faulting to parameters C_1, C_3, C_7, and C_8.

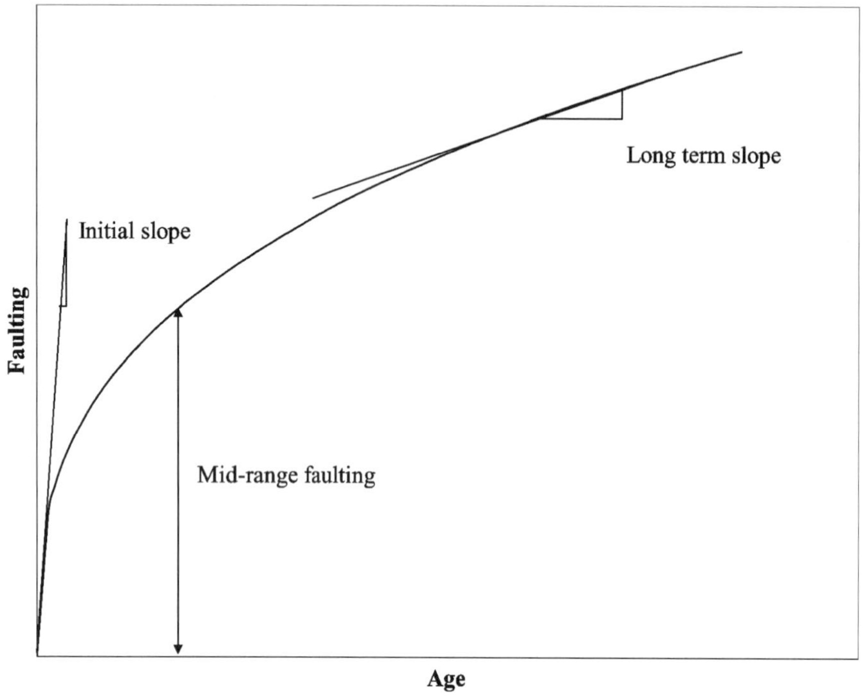

Figure A3-7. Different Stages of Faulting Development

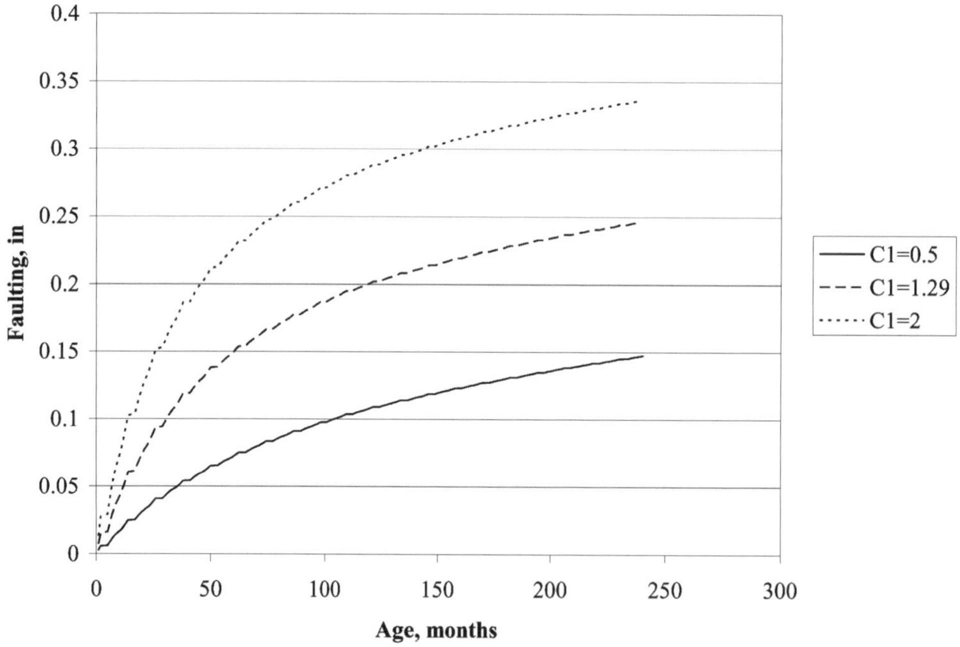

Figure A3-8. Effect of Parameter C_1 on Faulting Prediction

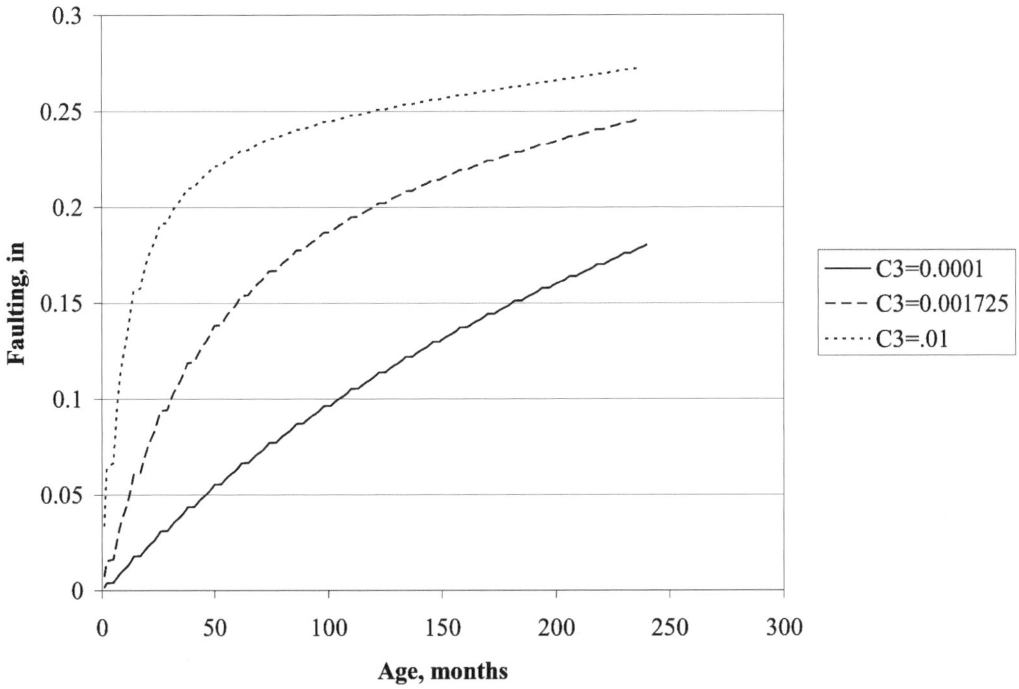

Figure A3-9. Effect of Parameter C_3 on Faulting Prediction

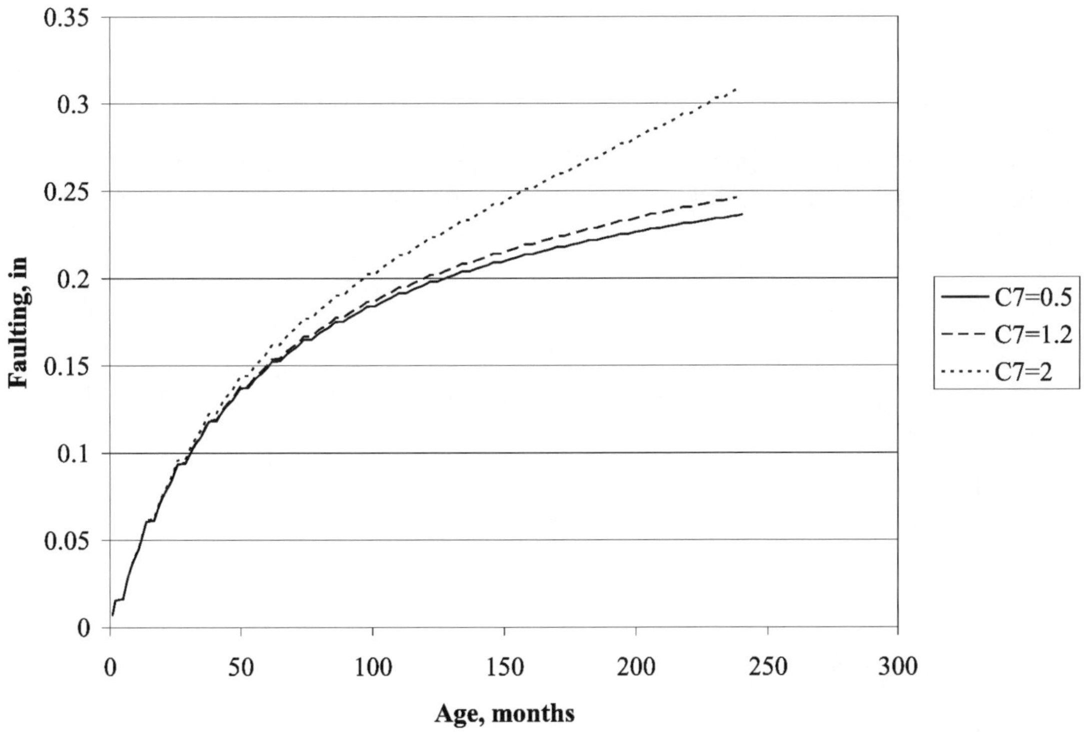

Figure A3-10. Effect of Parameter C_7 on Faulting Prediction

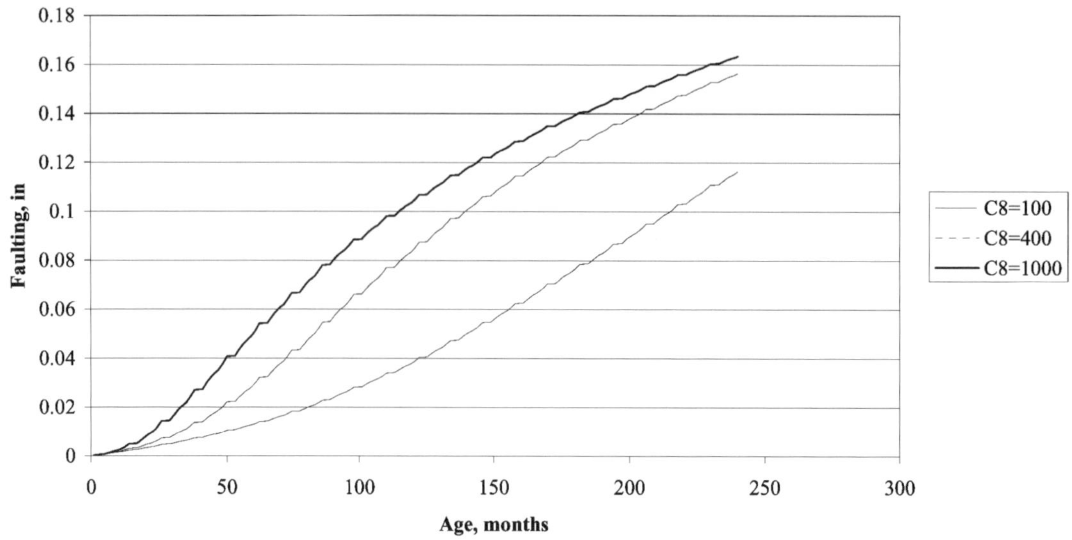

Figure A3-11. Effect of Parameter C_8 on Faulting Prediction

It can be observed that change of these coefficients may significantly change the shape of the faulting development curve and the magnitude of the predicted faulting. It should be also noted that change in the initial slope naturally changes mid-range faulting predictions and vise versa.
Coefficients C_5, C_6, and C_8 are responsible for magnifying the effect of different design feature on faulting prediction. Coefficient C_5 correlates change in erodibility with change in predicted faulting. Coefficient C_6 correlates influence of overburden on subgrade, percent subgrade material passing No. 200 sieve, and average annual number of wet days with faulting potential. Coefficient C_8 is responsible for the rate of deterioration of doweled joints. Figure A3-11 presents sensitivity of faulting prediction to parameters C_8 for sections with input parameters summarized in Table A3-8 but dowel diameter equal to 1 in. As shown in the figure, increasing C_8 leads to decrease of the effect of dowels on predicted faulting.

Based on the sensitivities of the faulting model coefficients on predicted faulting and observed inadequacies in predicted faulting, the following approach was adopted for local calibration:

- Adjust coefficients C_1, C_2, and C_7 as needed to influence the magnitudes of mid-range and long-range faulting.
- Adjust coefficients C_3 and C_4 as needed to influence the magnitudes of initial faulting. Coefficient C_8 can also be adjusted as needed to decrease the impact of dowels as needed.

The local calibration values for the faulting model coefficients were estimated by making repeat runs of the MEPDG with varying C_1, C_2, C_3, C_4, C_7, and C_8 values. The new local calibration coefficients obtained are as presented in Table A3-9.

Table A3-9. Summary of New Local Coefficients for Faulting Model

Faulting Coefficients	Global	Local
C_1	1.0184	0.1
C_2	0.91656	0.91656
C_3	0.002185	0.001
C_4	0.000884	0.000884
C_5	250	250
C_6	0.4	0.4
C_7	1.83312	3.0
C_8	400	400

The following summarizes the results from the local validation-calibration process.

- Coefficient C_7 was increased to increase predicted long term faulting.
- Coefficient C_3 was decreased to decrease predicted faulting in the mid range.
- Coefficient C_1 was decreased to decrease initial predicted faulting.

Table A3-10 lists the faulting bias using the local calibration values listed above. As shown, the hypothesis is now accepted, and the statistical parameters indicate a more accurate and precise faulting prediction model for the MODOT LTPP and PMS segments (Table A3-8 compared to Table A3-10). Figure A3-12 compares the predicted and measured faulting using the local calibration values, and shows that there is a substantial improvement, as compared to the use of the global calibration values (refer to Figure A3-5).

Table A3-10. Summary of the Statistical Parameters—Global Calibration Values Used for Predicting Performance Indicators for the Missouri LTPP and PMS Sections

Performance Indicator	Project	Bias (p-value)	Standard Error	R^2 (See Note 1)	Hypothesis; H_o: $y_i - \hat{y}_i = 0$	Comment
Transverse cracking	New	0.5283	—	Poor	Accept	No bias
Transverse joint faulting	New	0.1828[2]	0.03 in.	0.68	Accept	No bias
IRI	New	0.1478	11.1 in./mi	0.85	Accept	No bias

Notes: 1. Poor means that the model did not explain variation in the measured data within and between the LTPP and PMS segments.
2. Performed for only segments with faulting greater than 0.03 in.

Residual Error = $y_i - \hat{y}_i$
y_i = Measured or Observed Value
$\hat{y}_i$ = Predicted Value

Transverse Cracking

Model prediction using global coefficients was found to be adequate.

Roughness or IRI

Model prediction using global coefficients was found to be adequate.

Step 9—Assess Standard Error of the Estimate

After the bias was eliminated for each of the distress prediction models, the standard error of the estimate (refer to Table A3-10) is evaluated. For the runs using the local calibration values, the standard error was found to be statistically the same as the standard error included in the MEPDG for each performance indicator or there were too few segments with appreciable distress levels. Thus, the standard error of the estimate included in the MPEDG for each performance indicator should not be changed or altered.

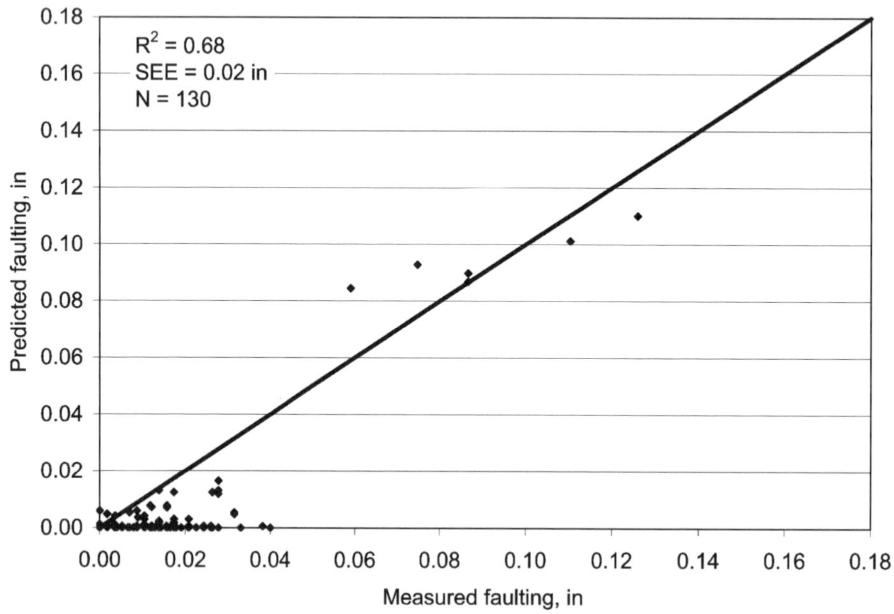

Figure A3-12. Comparison of Predicted and Measured Faulting Using the New Faulting Local Calibration Values for MODOT LTPP and PMS Segments

Step 10—Reduce Standard Error of the Estimate

As noted in Step 9, the standard error of the estimate was found to be statistically the same as the value reported in the MEPDG for each performance indicator.

Step 11—Interpretation of Results and Deciding on Adequacy of Calibration Factors

For this demonstration, the global calibration values did result in a bias for faulting distress. The MEPDG did not accurately predict faulting for JPCPs without dowels or with long (>20-ft) joint spacings. To reduce that bias required local calibration of the global faulting model.

The MEPDG transverse cracking and IRI models were confirmed using data from MODOT LTPP and PMS segments. The purpose of this step is to decide whether to adopt the locally calibrated faulting model or continue to using the global model (calibrated using data from LTPP projects from around the United States).

To make that decision, an agency should identify the major differences between the LTPP projects used in global calibration and the standard practice of the agency to specify, construct, and operate their roadway network. More importantly, the agency should determine whether the local calibration values can explain those differences. In addition, the agency should evaluate the new locally calibrated model to establish its engineering reasonableness by performing a comprehensive sensitivity analysis.

A3.2 Attachments

A3.2.A Attachment A—Description of MODOT LTPP and PMS JPCP Segments

A3.2.A.1 Design (Analysis) Life

The MEPDG requires pavement construction and traffic opening dates along with design life or analysis period. Design life for each project was determined based on construction date. A summary of pavement construction dates and analysis life is presented in Table A3-11.

Table A3-11. Summary of Construction Dates and Analysis Periods for All New JPCP

Section ID	Design Life, yrs	Construction Date	Traffic Open Date	Pavement Type
29_0701_1	53	September 1955	October 1955	JPCP
29_0807_1	10	May 1998	June 1998	JPCP
29_0808_1	10	May 1998	June 1998	JPCP
29_5393_1	51	October 1957	November 1957	JPCP
29_A601b	40	July 1969	August 1969	JPCP
29_A807_1	10	November 1998	December 1998	JPCP
29_A808_1	10	November 1998	December 1998	JPCP
A01-W-S1	19	July 1989	August 1989	JPCP
A01-W-S2	19	July 1989	August 1989	JPCP
A02-E-S1	21	July 1987	August 1987	JPCP
A02-W-S1	21	July 1987	August 1987	JPCP
A03-W-S1	33	September 1975	October 1975	JPCP
A03-W-S2	33	September 1975	October 1975	JPCP
A03-W-S3	33	September 1975	October 1975	JPCP
B01-W-S1	13	July 1995	August 1995	JPCP
C01-W-S1	11	July 1997	August 1997	JPCP
D01-W-S1	13	September 1995	October 1995	JPCP
D03-E-S1	12	September 1996	October 1996	JPCP
E01-S-S1	12	October 1996	November 1996	JPCP
E01-S-S2	12	October 1996	November 1996	JPCP
E01-S-S3	12	October 1996	November 1996	JPCP
E02-S-S1	13	August 1995	September 1995	JPCP
E02-S-S2	13	August 1995	September 1995	JPCP
E04-S-S1	11	September 1997	October 1997	JPCP
E04-S-S2	11	September 1997	October 1997	JPCP
F01-W-S1	12	October 1996	November 1996	JPCP

Table A3-11—Continued

Section ID	Design Life, yrs	Construction Date	Traffic Open Date	Pavement Type
F01-W-S2	12	October 1996	November 1996	JPCP
F01-W-S3	12	October 1996	November 1996	JPCP
F01-W-S4	12	October 1996	November 1996	JPCP
F02-E-S1	14	July 1994	August 1994	JPCP
F03-W-S1	14	July 1994	August 1994	JPCP
F03-W-S2	14	August 1994	September 1994	JPCP
F03-W-S3	14	August 1994	September 1994	JPCP
F04-W-S1	14	June 1994	July 1994	JPCP
F05-E-S1	15	July 1993	August 1993	JPCP
F05-E-S2	15	July 1993	August 1993	JPCP
F05-E-S3	15	July 1993	August 1993	JPCP
F06-E-S1	14	September 1994	October 1994	JPCP
F06-E-S2	14	September 1994	October 1994	JPCP
F06-E-S3	14	September 1994	October 1994	JPCP
F07-S-S1	11	July 1997	August 1997	JPCP
F07-S-S2	11	July 1997	August 1997	JPCP
F08-E-S1	11	June 1997	July 1997	JPCP
F08-E-S2	11	June 1997	July 1997	JPCP
F09-W-S1	11	July 1997	August 1997	JPCP
F09-W-S2	11	July 1997	August 1997	JPCP
F10-E-S1	13	September 1995	October 1995	JPCP
G01-S-S1	14	October 1994	November 1994	JPCP
G01-S-S2	14	October 1994	November 1994	JPCP
G02-E-S1	10	April 1998	May 1998	JPCP
G02-E-S2	10	April 1998	May 1998	JPCP
G03-E-S1	10	September 1998	October 1998	JPCP
G03-E-S2	10	September 1998	October 1998	JPCP
G03-E-S3	10	September 1998	October 1998	JPCP

A3.2.A.2 Analysis Parameters

The MEPDG requires terminal distress/IRI values along with initial IRI. For the validation exercise, terminal distress/IRI is not relevant. For all the projects used in analysis, initial IRI was backcast from historical IRI data available for each section. Although there was a wide range of backcast initial IRI values, the estimated initial IRI was deemed reasonable, with values shown in Table A3-12.

Table A3-12. Summary of Backcast Initial IRI Values

Pavement Type	IRI, in./mi		
	Minimum	Maximum	Average
JPCP	34	158*	73.1

A3.2.A.3 Traffic

Many of the traffic inputs were obtained at Level 1 or 2, since weigh-in-motion (WIM) and automated vehicle classification (AVC) data were available for all the projects (see Figure A3-13). The majority of WIM sites were situated on rural interstate or primary highways. The specific routes with WIM sites are as follows:

- Interstate: I-35 (rural), I-29 (rural), I-55 (rural), I-65 (rural), I-70 (rural), I-44 (rural), I-71 (rural), I-435 (urban), I-635 (urban).
- US/Primary: US 61, US 65, US 54, US 75, US 60, US 40, US 412—all rural.
- Other: Route C, MO 171, MO 210.

For most of the WIM sites, there were one to seven years of data available, as shown in Figure A3-14. The traffic data obtained from the WIM sites were analyzed to develop MEPDG traffic inputs. Traffic data processing consisted of the following steps:

1. Rate traffic data quality. Quality rating was based on consistency of multi-year class and load spectra information. The rating scale (after presumed outliers removed) was as follows:
   5—Very good; consistent and clean trends.
   4—Good; reasonable, consistent trends.
   3—Average; reasonable trends after considerable filtering.
   2—Poor; data questionable, some misclassifications.
   1—Very poor; significant amount of data questionable, several misclassifications.

2. Traffic data rated as poor or very poor were included in the database. Note that 16 WIM sites were rated as having reasonable quality data while 8 WIM sites had questionable data. Determine the representative MEPDG Truck Traffic Classification (TTC) grouping for each WIM site (e.g., TTC 1).

3. Using the WIM data assembled, the following MEPDG traffic inputs were estimated for each TTC group:
   - Hourly distribution of traffic.
   - Vehicle class distribution (see Figure A3-15).
   - Number of axles per truck (see Figures A3-16 and A3-17).
   - Axle load distribution (see Figure A3-18).

4. Individual project vehicle counts and percent trucks (not always from the WIM sites) were used to determine both initial AADTT and truck traffic growth rate and type.

A summary of MEPDG traffic volume inputs is presented in Table A3-13.

Table A3-13. Summary of Traffic Inputs for All New JPCP Projects

Section ID	Initial Two-Way AADTT	Number Lanes in Design Direction	Percent of Trucks in Design Direction, %	Percent of Trucks in Design Lane, %	Operational Speed, mph
29_0701_1	244	1	100	100	60
29_0807_1	80	1	100	100	60
29_0808_1	80	1	100	100	60
29_5393_1	117	1	100	100	60
29_A601b	100	1	100	100	60
29_A807_1	180	1	100	100	60
29_A808_1	180	1	100	100	60
A01-W-S1	81	1	50	100	60
A01-W-S2	81	1	50	100	60
A02-E-S1	20	1	50	100	60
A02-W-S1	20	1	50	100	60
A03-W-S1	953	2	50	95	60
A03-W-S2	953	2	50	95	60
A03-W-S3	953	2	50	95	60
B01-W-S1	198	1	50	100	60
C01-W-S1	25	1	50	100	60
D01-W-S1	462	2	50	95	60
D03-E-S1	330	1	50	100	60
E01-S-S1	2357	2	50	95	60
E01-S-S2	2357	2	50	95	60
E01-S-S3	2357	2	50	95	60
E02-S-S1	2302	2	50	95	60
E02-S-S2	2302	2	50	95	60
E04-S-S1	1399	2	50	95	60
E04-S-S2	1399	2	50	95	60
F01-W-S1	1245	2	50	95	60
F01-W-S2	1245	2	50	95	60
F01-W-S3	1245	2	50	95	60
F01-W-S4	1245	2	50	95	60
F02-E-S1	973	2	50	95	60
F03-W-S1	856	2	50	95	60
F03-W-S2	856	2	50	95	60
F03-W-S3	856	2	50	95	60
F04-W-S1	1020	2	50	95	60
F05-E-S1	903	2	50	95	60
F05-E-S2	903	2	50	95	60
F05-E-S3	903	2	50	95	60
F06-E-S1	1095	2	50	95	60
F06-E-S2	1095	2	50	95	60

Table A3-13—Continued

Section ID	Initial Two-Way AADTT	Number Lanes in Design Direction	Percent of Trucks in Design Direction, %	Percent of Trucks in Design Lane, %	Operational Speed, mph
F06-E-S3	1095	2	50	95	60
F06-E-S1	1095	2	50	95	60
F06-E-S2	1095	2	50	95	60
F06-E-S3	1095	2	50	95	60
F07-S-S1	1640	2	50	95	60
F07-S-S2	1640	2	50	95	60
F08-E-S1	833	2	50	95	60
F08-E-S2	833	2	50	95	60
F09-W-S1	1631	2	50	95	60
F09-W-S2	1631	2	50	95	60
F10-E-S1	586	2	50	95	60
G01-S-S1	260	2	50	95	60
G01-S-S2	260	2	50	95	60
G02-E-S1	496	2	50	95	60
G02-E-S2	496	2	50	95	60
G03-E-S1	421	2	50	95	60
G03-E-S2	421	2	50	95	60
G03-E-S3	421	2	50	95	60

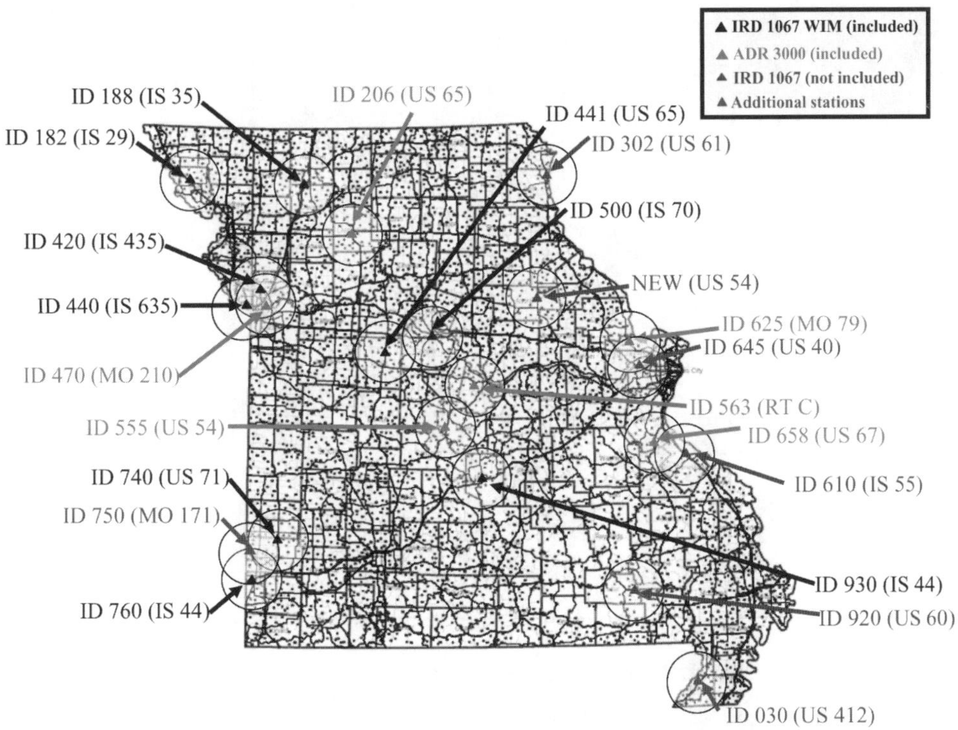

Figure A3-13. Locations of WIM Sites in Missouri from Which Traffic Data Were Obtained for Validation-Recalibration

Table A3-13—Continued

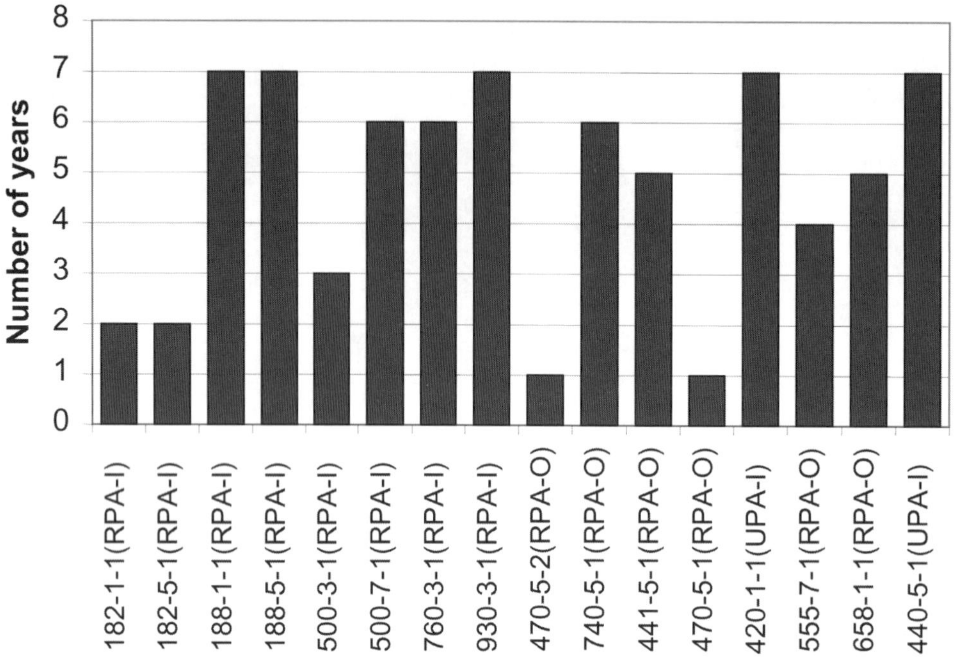

Figure A3-14. Summary of Years of WIM Data Available

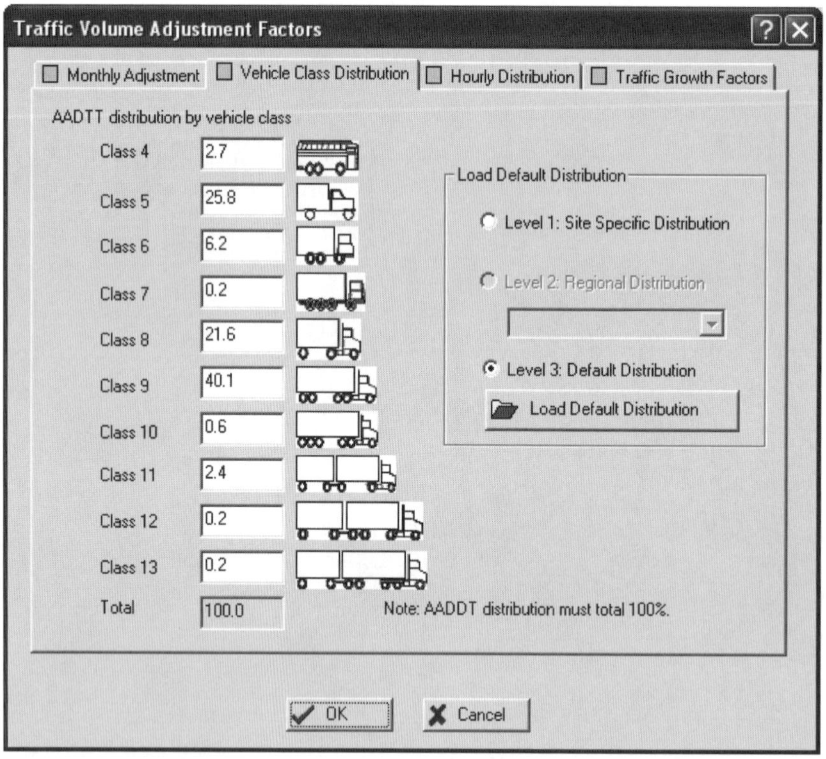

Figure A3-15. Monthly Truck Volume Adjustment Factors for 29-1005

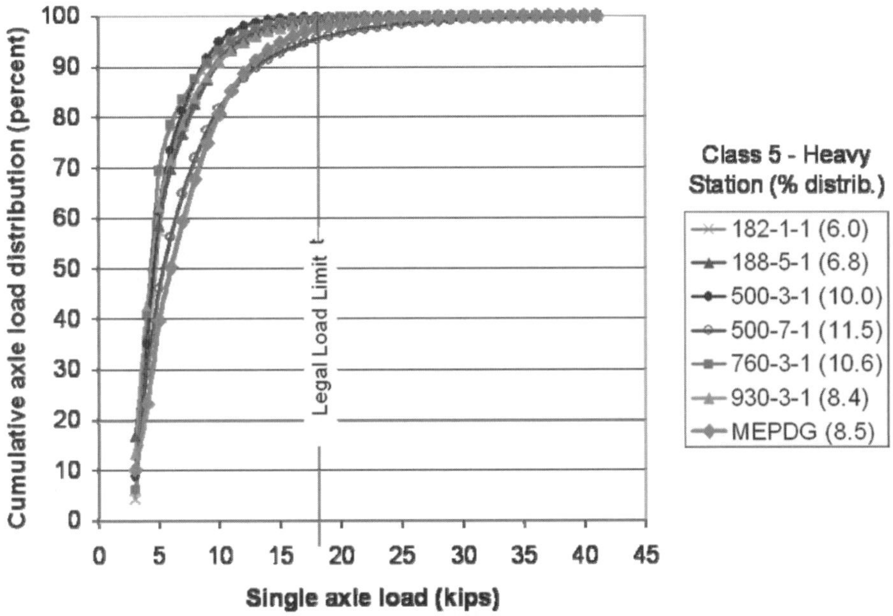

Figure A3-16. Cumulative Single-Axle-Load Distribution for Class 5 Trucks on MODOT Heavy-Duty Pavements Pertaining to TTC 1

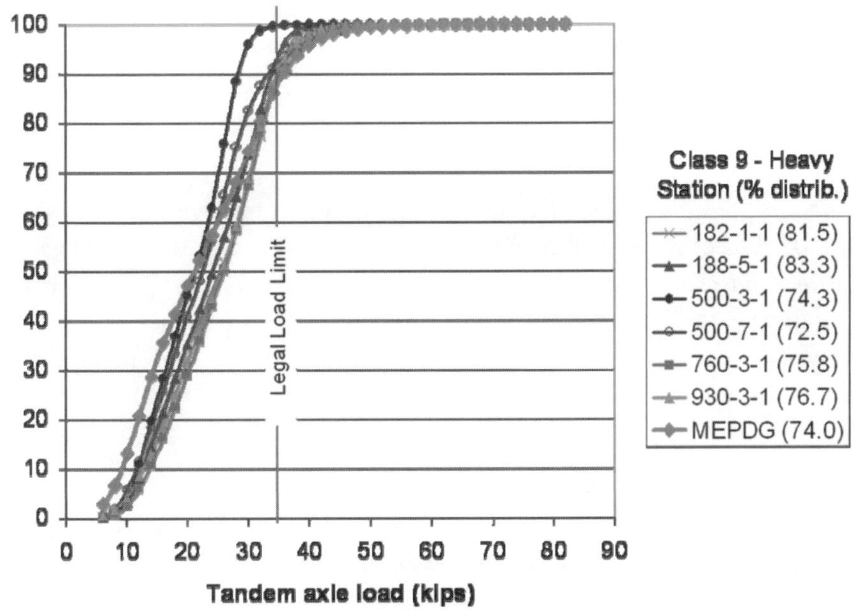

Figure A3-17. Cumulative Tandem Axle-Load Distribution for Class 9 Trucks on MODOT Heavy-Duty Pavements Pertaining to TTC 1

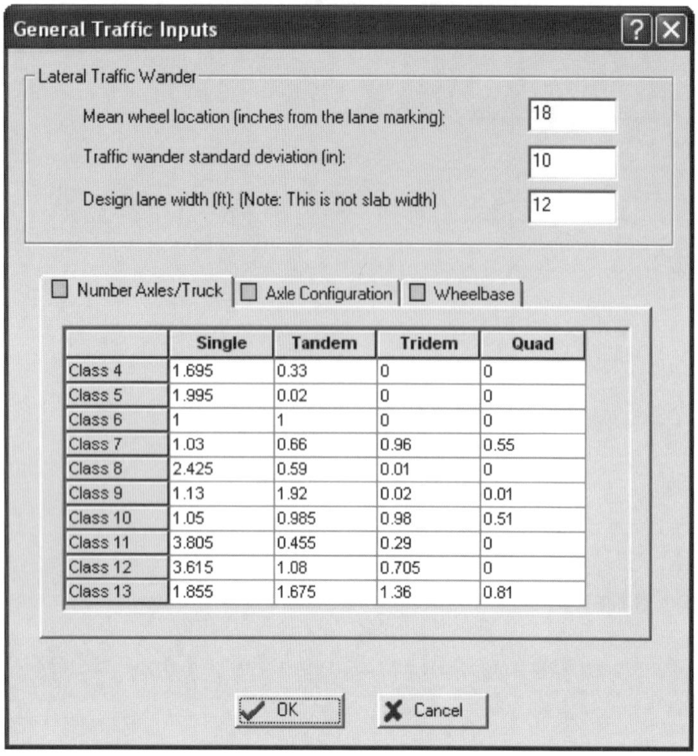

Figure A3-18. Lateral Truck Wander and Mean Number Axles/Truck for 29-1005

A3.2.A.4 Climate

The MEPDG requires the location of a project described in terms of longitude, latitude, and elevation in order to develop project-specific climate data for analysis. The climate data for each project were generated using up to six of the closest weather stations. Typically, each weather station had 96 to 116 months of climate data. Another piece of information that is required along with project location information is an estimate of depth-to-water table level. For this project, a default depth to water table ranging from 3 to 25 ft was adopted (based on local Missouri conditions).

A3.2.A.5 Pavement Surface Layer Thermal Properties

The MEPDG default surface shortwave absorptivity, thermal conductivity, and heat capacity were used for all the layers and for the analyses performed.

A3.2.A.6 Design Features for JPCP Sections

The following JPCP design features are required:

- The temperature gradient during PCC placement and curing.
- PCC slab transverse joint spacing.
- Transverse joint sealant type.
- Slab width.
- Load transfer mechanism and properties.
- Slab edge support type.
- Base type and base erosion factor.
- PCC-base interface friction type and age at which friction is lost.

Details are presented in Table A3-14.

Table A3-14. JPCP Project Design Features

Project ID	Joint Spacing, ft	PCC Slab Edge Support		Transverse Joint Load Transfer	
		Slab Width, ft	Tied PCC Shoulder (Y/N)	Dowel Diameter, in.	Dowel Spacing, in.
29_0701_1	20	12	No	0	0
29_0807_1	15	10	No	1.25	12
29_0808_1	15	10	No	1.25	12
29_5393_1	30	12	No	1.25	12
29_A601b	30	12	No	1	15
29_A807_1	15	10	No	1.25	12
29_A808_1	15	10	No	1.25	12
A01-W-S1	30	12	No	1.25	12
A01-W-S2	30	12	No	1.25	12
A02-E-S1	30	12	No	1.25	12
A02-W-S1	30	12	No	1.25	12
A03-W-S1	30	12	No	1.25	12
A03-W-S2	30	12	No	1.25	12
A03-W-S3	30	12	No	1.25	12
B01-W-S1	15	12	No	1.25	12
C01-W-S1	15	14	No	1.25	12
D01-W-S1	15	12	Yes	1.25	12
D03-E-S1	15	14	Yes	1.25	12
E01-S-S1	15	12	Yes	1.5	12
E01-S-S2	15	12	Yes	1.5	12
E01-S-S3	15	12	Yes	1.5	12
E02-S-S1	15	12	Yes	1.5	12
E02-S-S2	15	12	Yes	1.5	12
E04-S-S1	15	12	Yes	1.5	12
E04-S-S2	15	12	Yes	1.5	12
F01-W-S1	15	12	Yes	1.5	12
F01-W-S2	15	12	Yes	1.5	12
F01-W-S3	15	12	Yes	1.5	12
F01-W-S4	15	12	Yes	1.5	12
F02-E-S1	15	12	Yes	1.5	12
F03-W-S1	15	12	Yes	1.5	12
F03-W-S2	15	12	Yes	1.5	12
F03-W-S3	15	12	Yes	1.5	12
F04-W-S1	15	12	Yes	1.5	12
F05-E-S1	15	12	Yes	1.5	12
F05-E-S2	15	12	Yes	1.5	12
F05-E-S3	15	12	Yes	1.5	12
F06-E-S1	15	12	Yes	1.5	12

Table A3-14—Continued

Project ID	Joint Spacing, ft	PCC Slab Edge Support		Transverse Joint Load Transfer	
		Slab Width, ft	Tied PCC Shoulder (Y/N)	Dowel Diameter, in.	Dowel Spacing, in.
F06-E-S2	15	12	Yes	1.5	12
F06-E-S3	15	12	Yes	1.5	12
F07-S-S1	15	12	Yes	1.5	12
F07-S-S2	15	12	Yes	1.5	12
F08-E-S1	15	12	Yes	1.5	12
F08-E-S2	15	12	Yes	1.5	12
F09-W-S1	15	12	Yes	1.5	12
F09-W-S2	15	12	Yes	1.5	12
F10-E-S1	15	12	Yes	1.5	12
G01-S-S1	15	12	Yes	1.5	12
G01-S-S2	15	12	Yes	1.5	12
G02-E-S1	15	12	Yes	1.5	12
G02-E-S2	15	12	Yes	1.5	12
G03-E-S1	15	12	Yes	1.5	12
G03-E-S2	15	12	Yes	1.5	12
G03-E-S3	15	12	Yes	1.5	12
29_5393_1	30	12	No	1.25	12

- Default permanent curl/warp temperature of −10°F was assumed.
- Default liquid joint sealant was assumed.
- Full friction between the PCC slab and underlying base layer was assumed.

A3.2.A.7 Pavement Structure Definition

The MEPDG requires a definition of the pavement structure along with a detailed description/characterization of the layer materials that make up the pavement structure. Pavement structure is defined by layer material type, position within the structure, and thickness. Material characterization mostly consists of properties needed to support climate modeling, response analysis, and performances prediction.

For all the material groups, detailed information was obtained from the MODOT and LTPP database and used to characterize the layer material properties including thickness, unit weight, Poisson's ratio, gradation, asphalt mix properties, PCC flexural strength, PCC thermal coefficient of expansion, and PCC modulus of elasticity. Most of the key material properties in the databases were obtained through laboratory testing of mix samples or extracted cores. For other material properties, such as PCC zero stress temperature, thermal conductivity, and so on, MEPDG or Missouri-specific defaults were assumed.

A3.2.B Attachment B—Plots of Time-History Performance Data

The following are plots of the time-history performance data that provide performance data over time for all of the MODOT-LTPP and PMS-JPCP segments. These time-history plots were used to determine if any anomalies or outliers were present in the data.

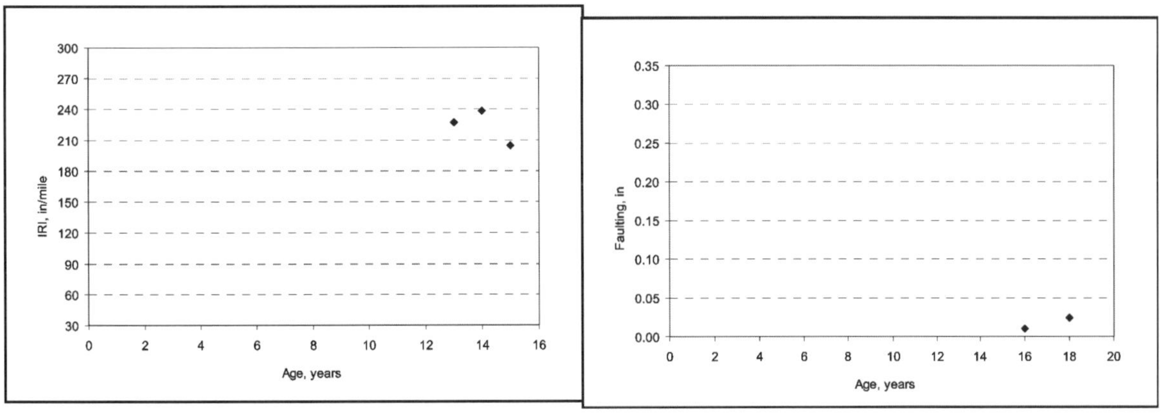

PMS site—A01-W-S1

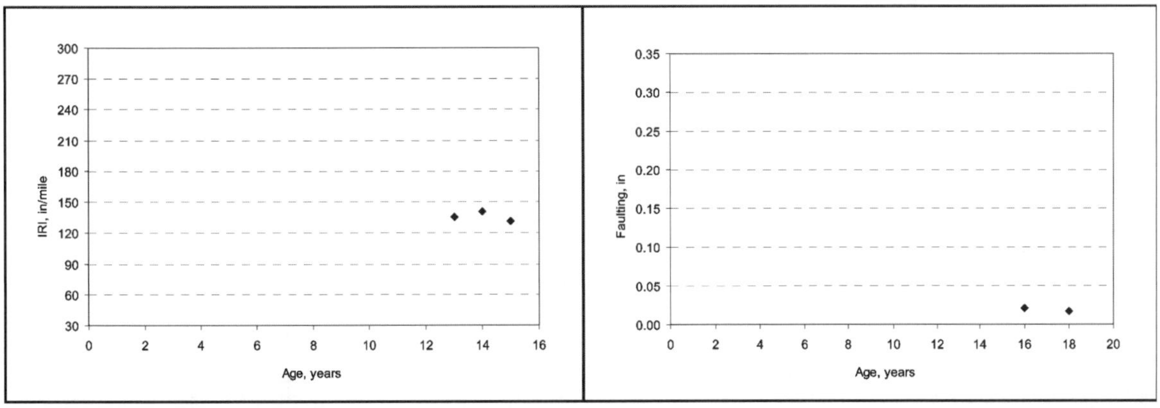

PMS site—A01-W-S2

Figure A3.2.B-1. PMS Sites

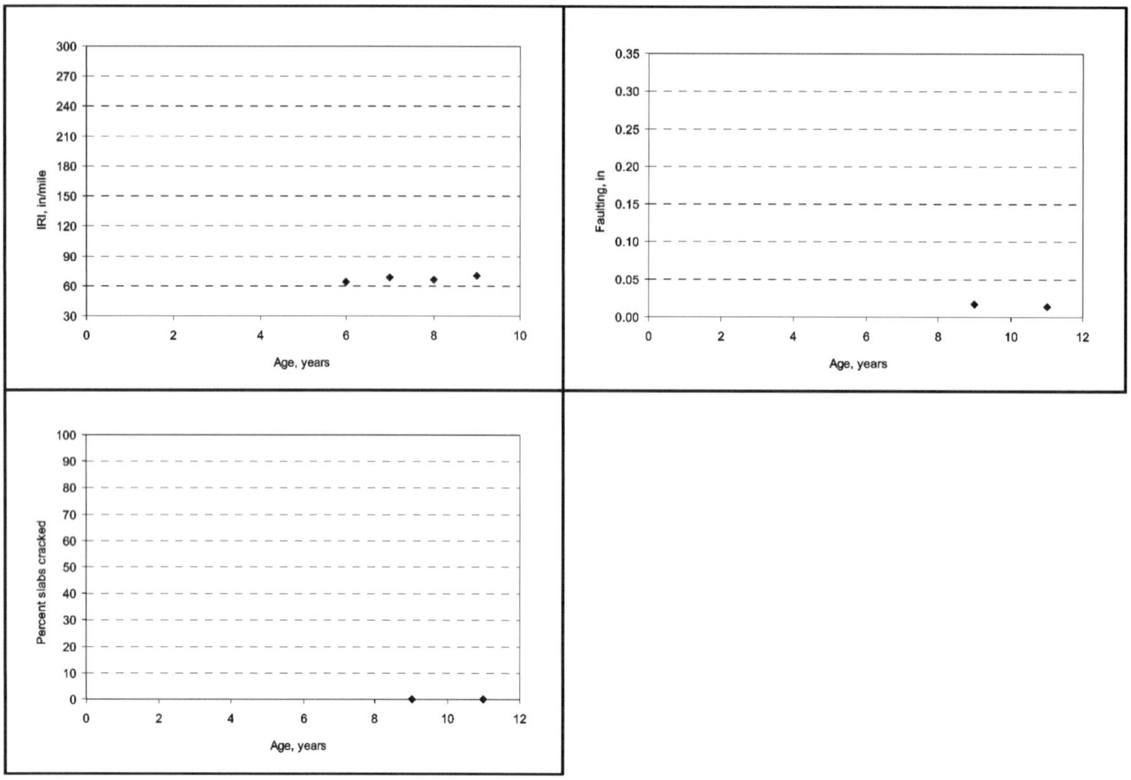

PMS site - F01-W-S2

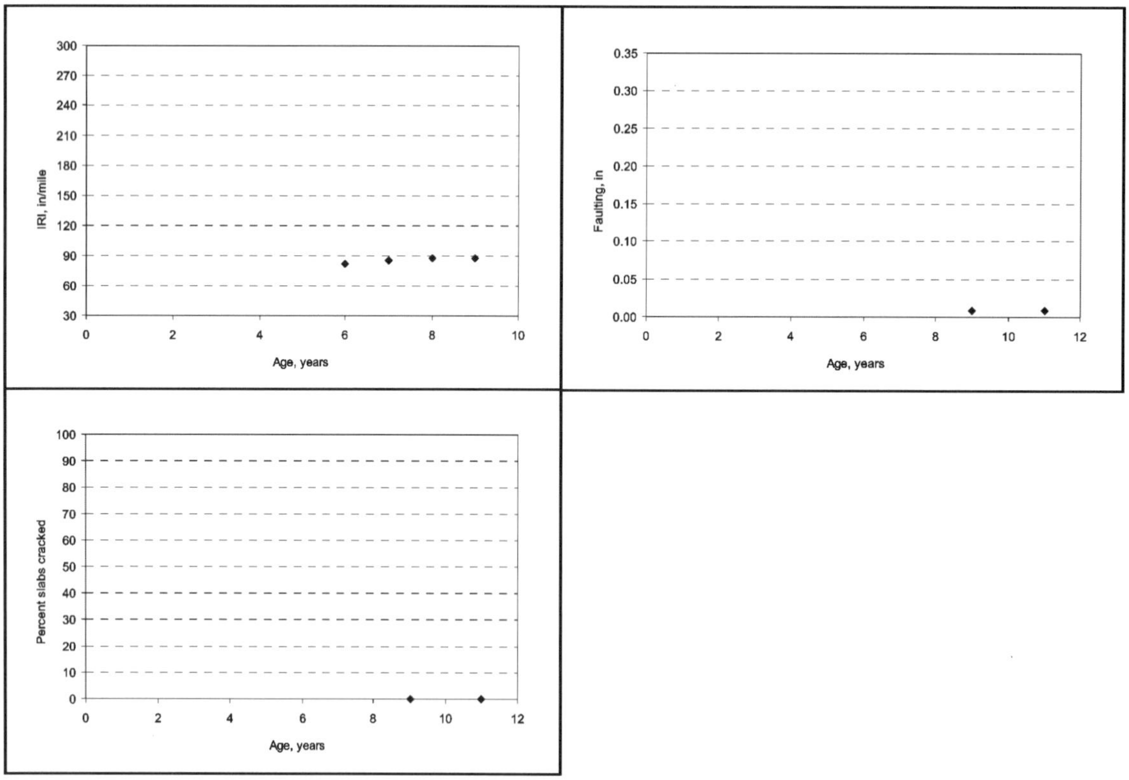

PMS site - F01-W-S3

Figure A3.2.B-1.—*Continued*

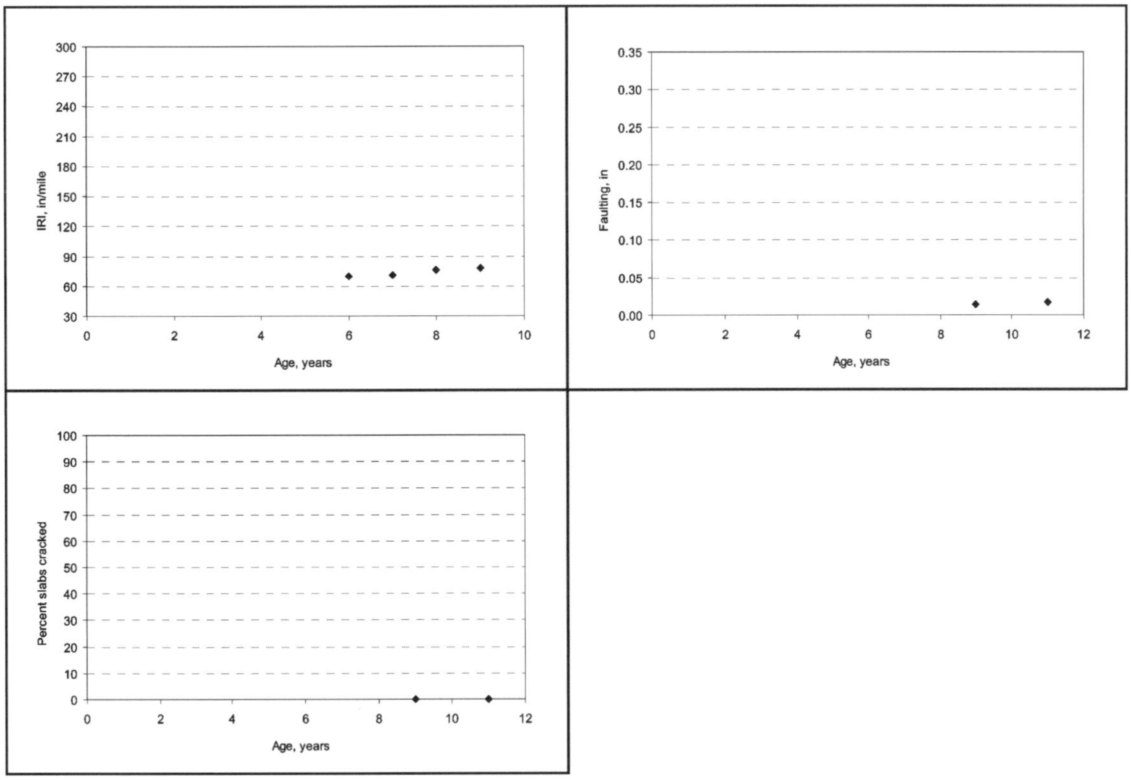

PMS site—F01-W-S4

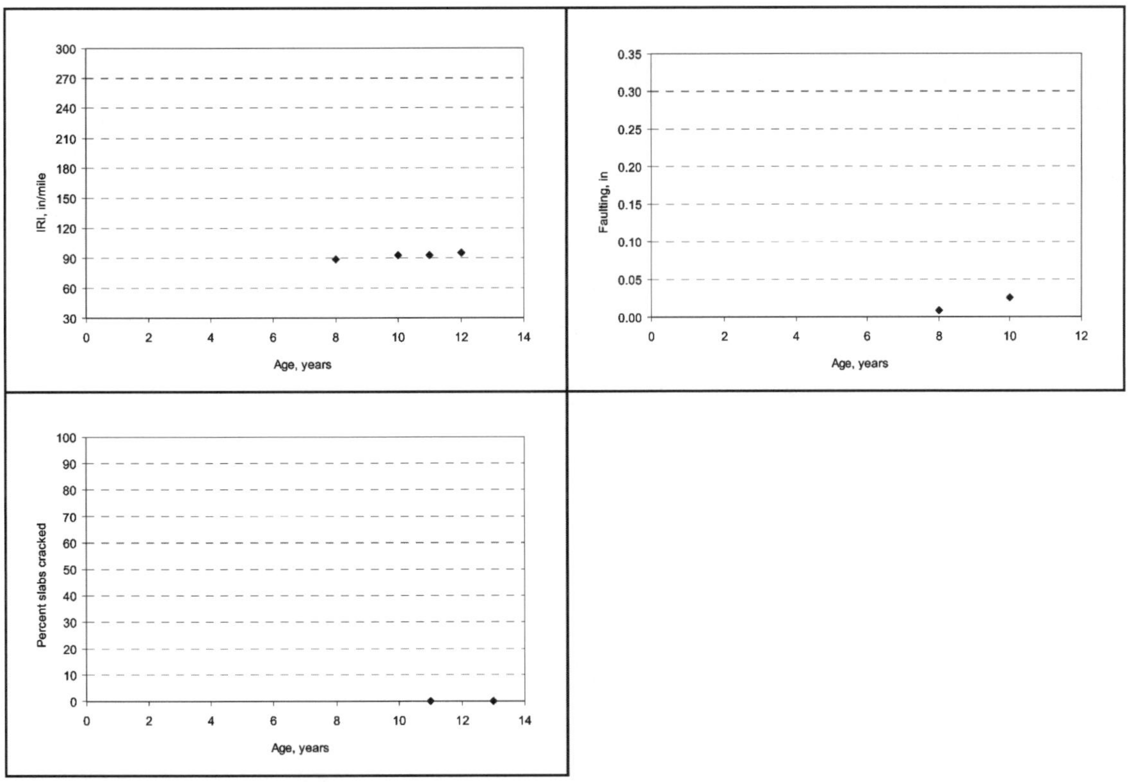

PMS site—F02-E-S1

Figure A3.2.B-1.—*Continued*

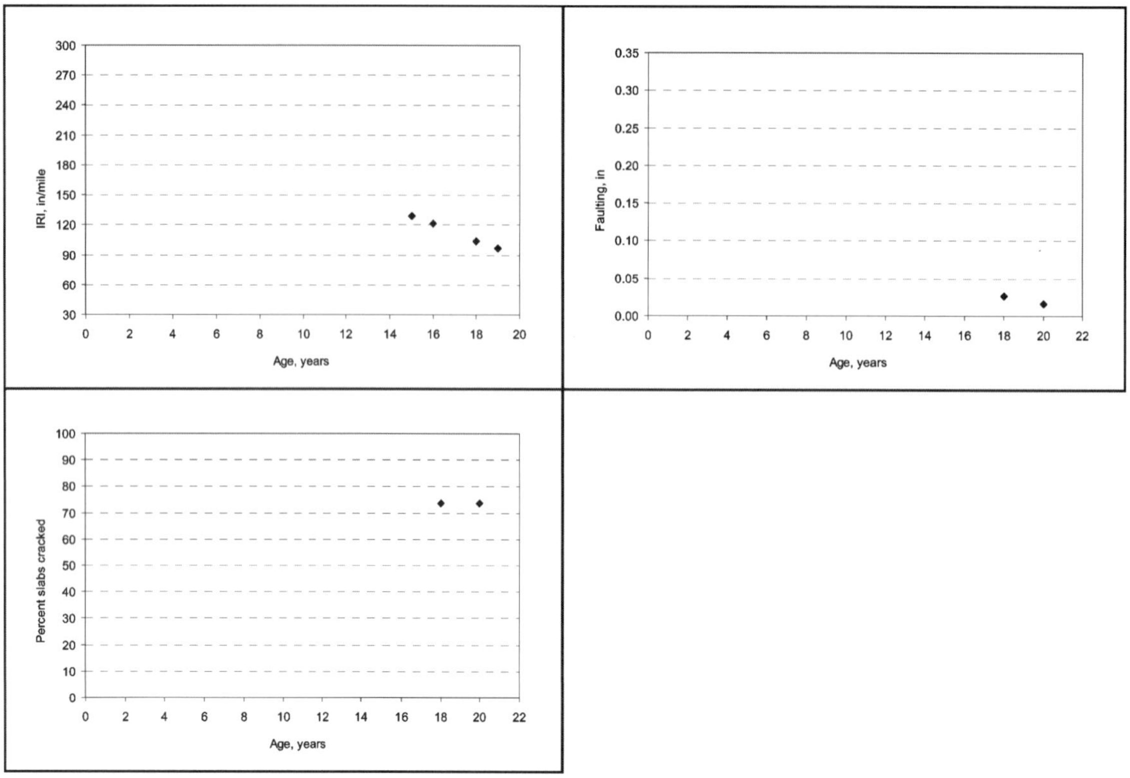

PMS site—A02-E-S1

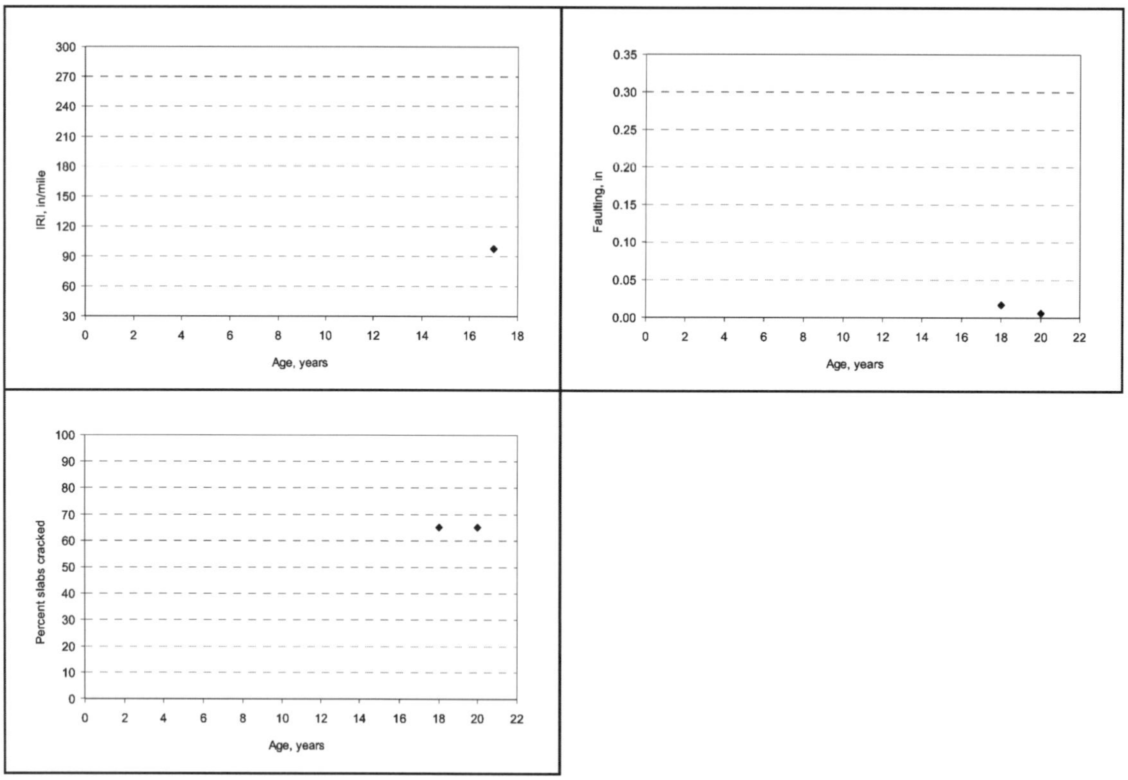

PMS site—A02-W-S1

Figure A3.2.B-1.—*Continued*

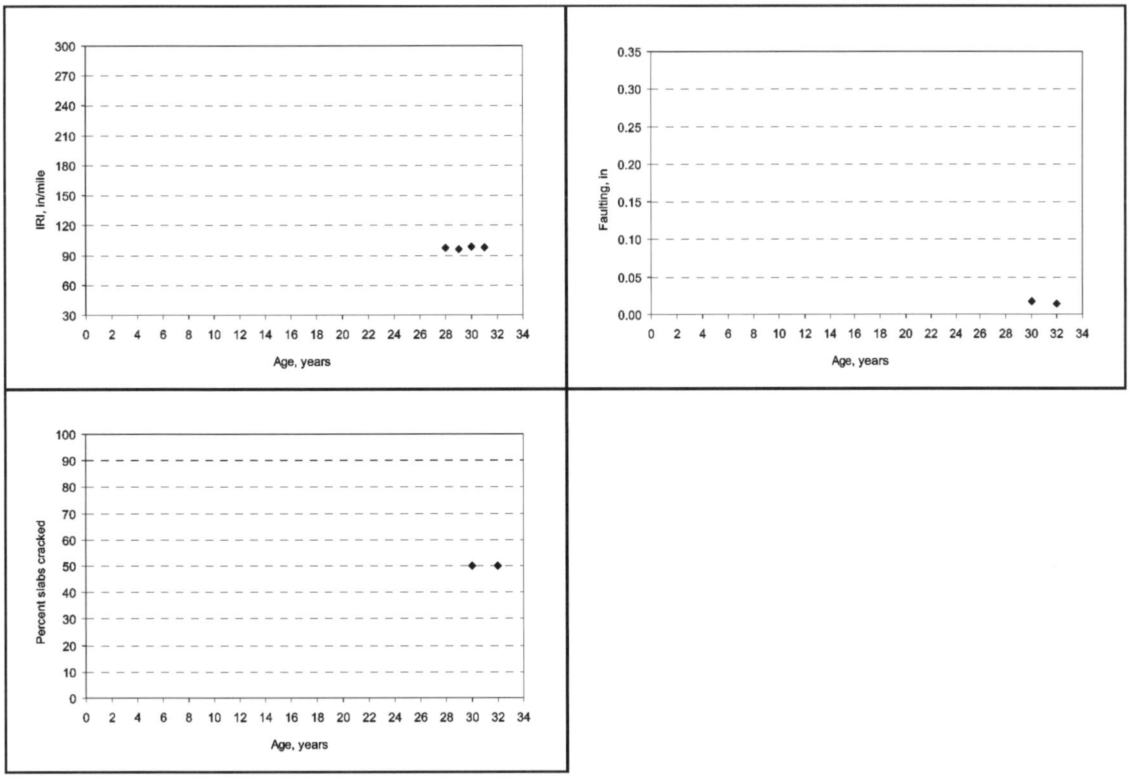

PMS site—A03-W-S1

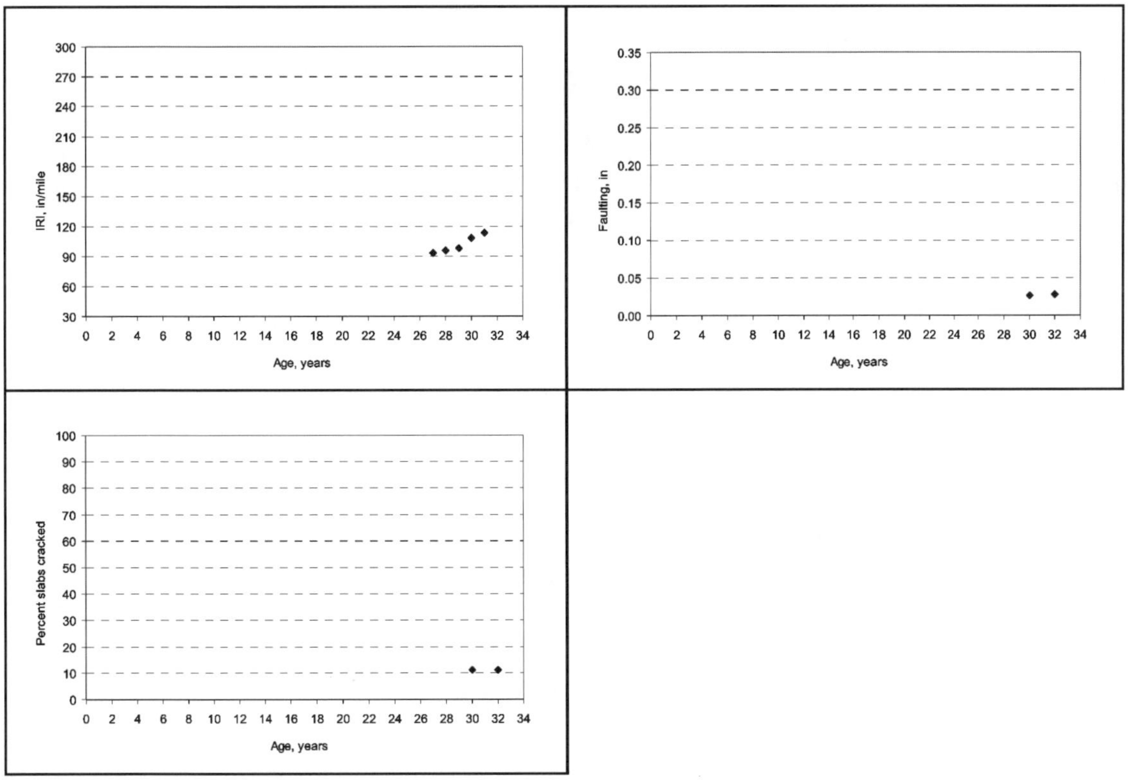

PMS site—A03-W-S2

Figure A3.2.B-1.—*Continued*

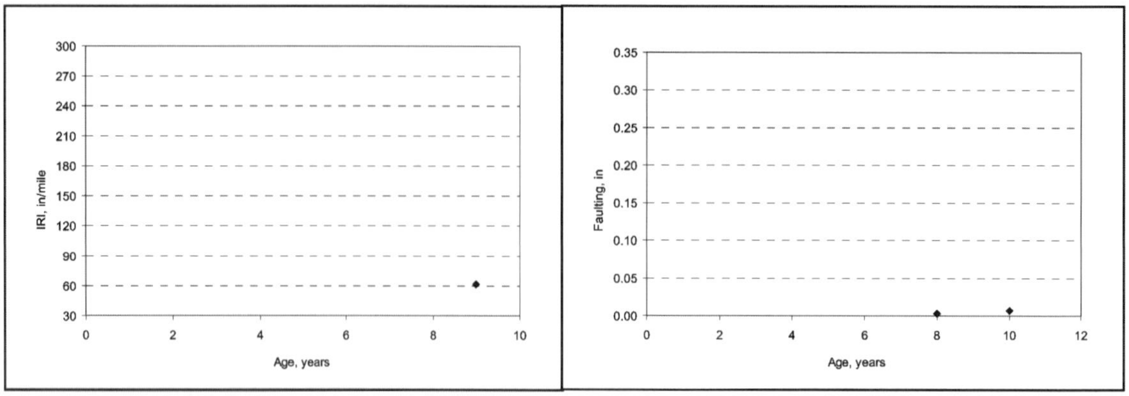

PMS site—C01-W-S1

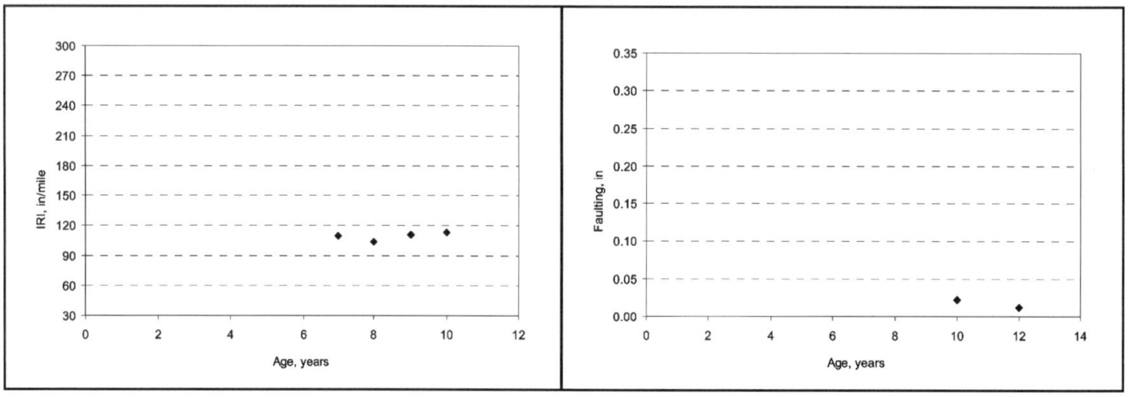

PMS site—D01-W-S1

Figure A3.2.B-1.—*Continued*

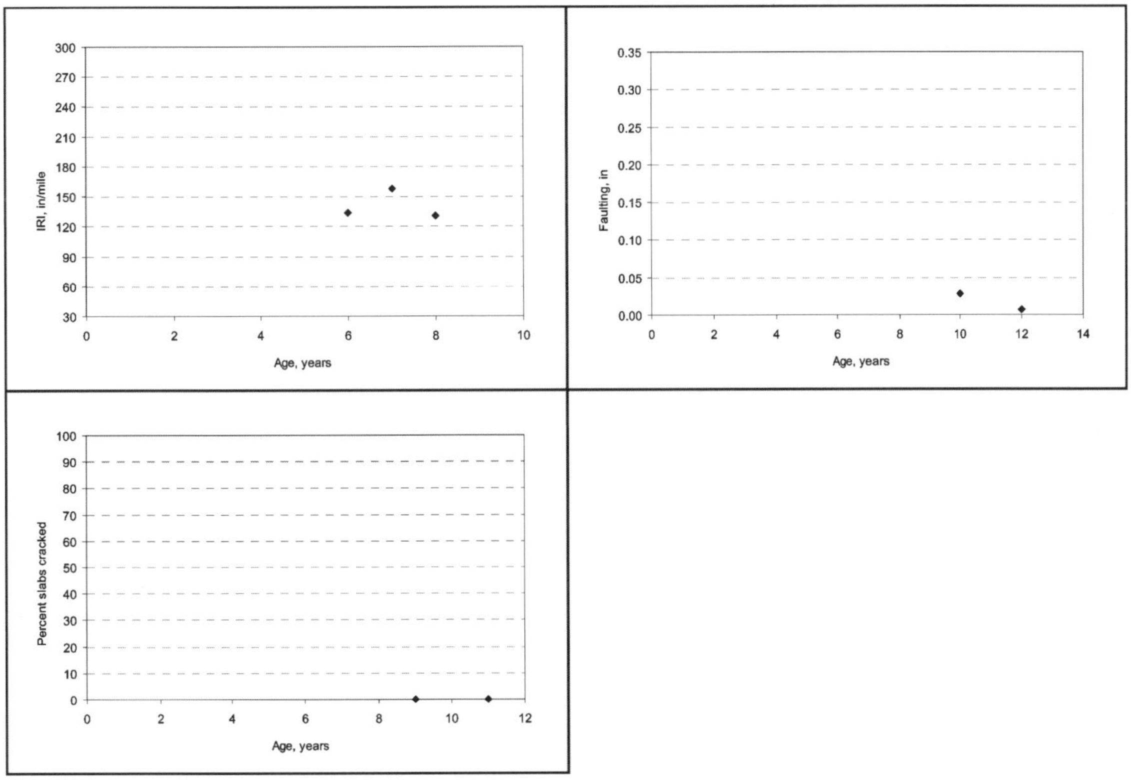

PMS site—D03-E-S1

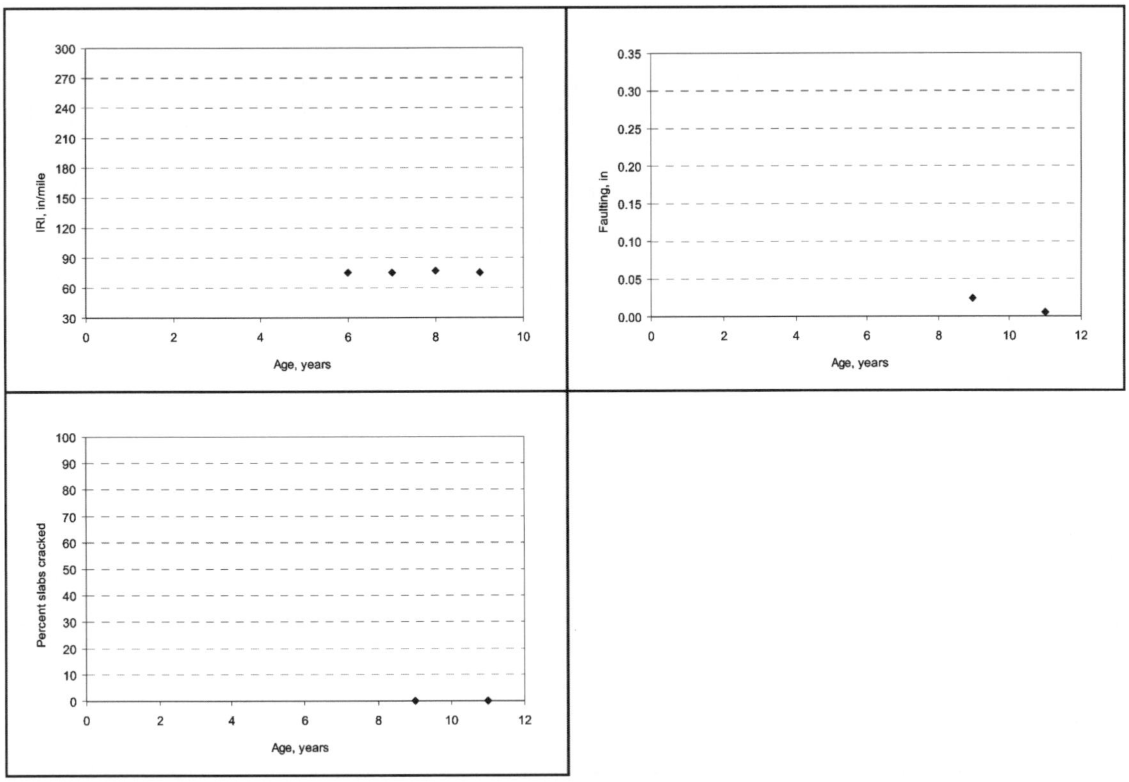

PMS site—E01-S-S1

Figure A3.2.B-1.—*Continued*

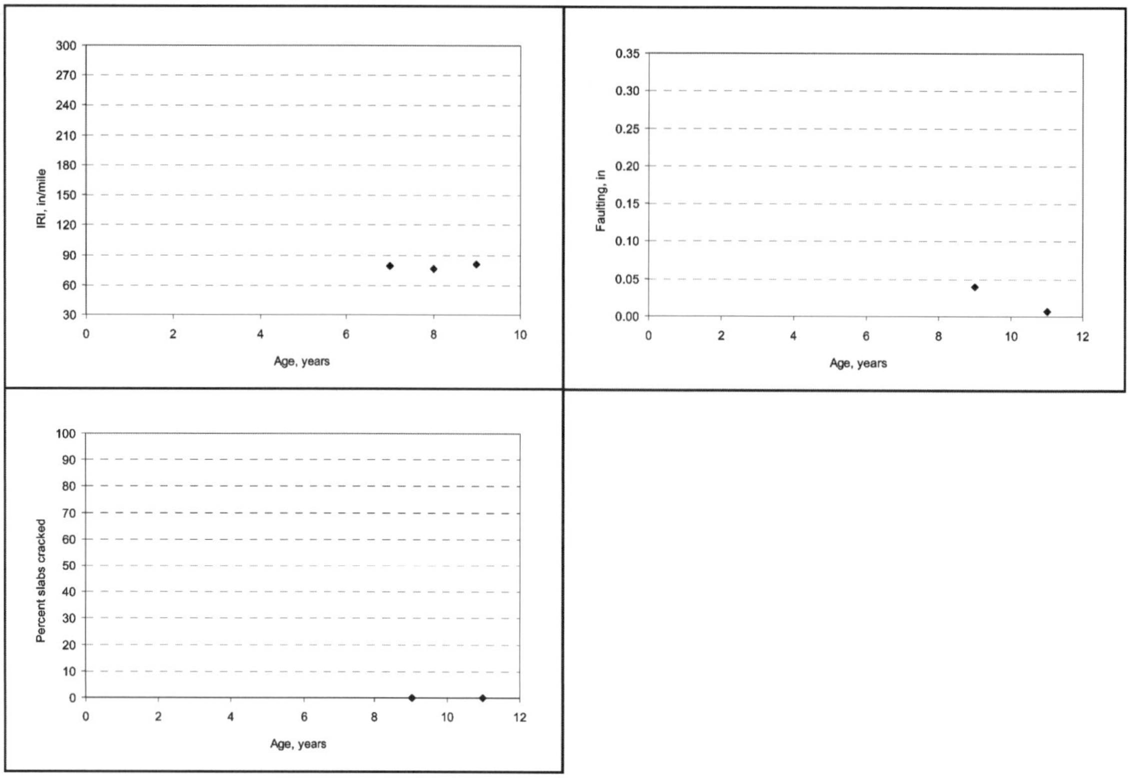

PMS site—E01-S-S2

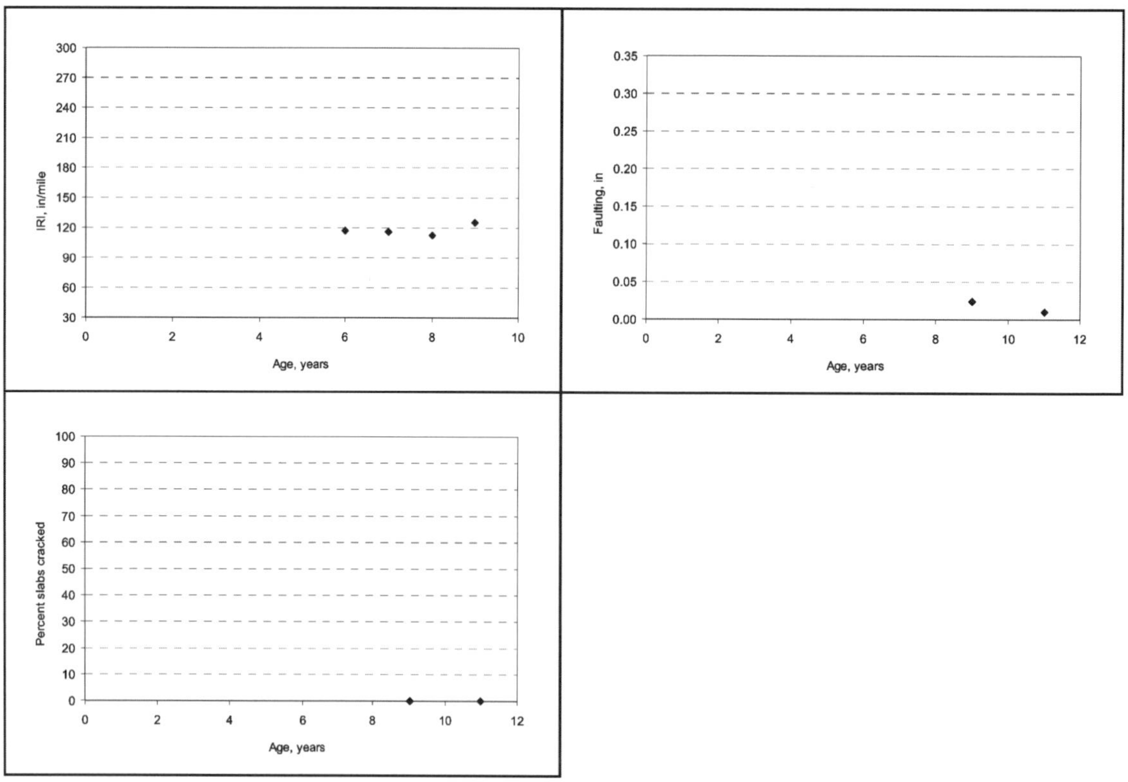

PMS site—E01-S-S3

Figure A3.2.B-1.—*Continued*

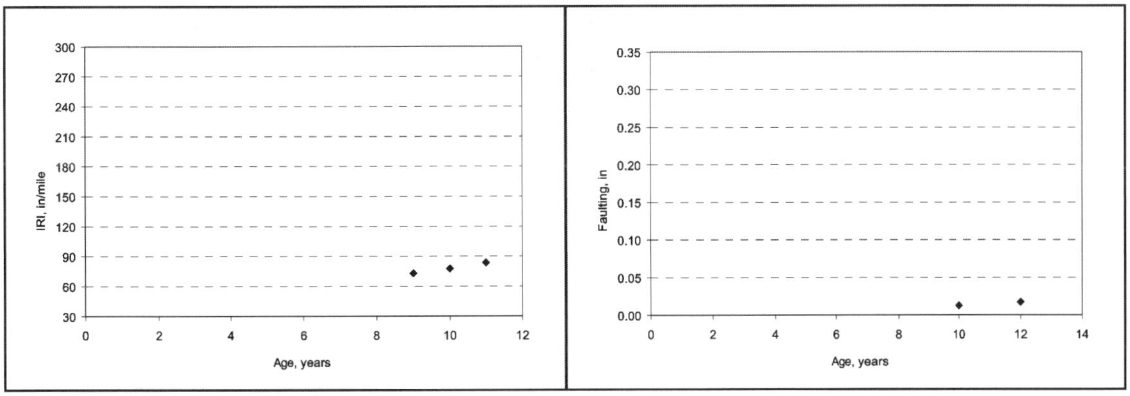

PMS site—E02-S-S1

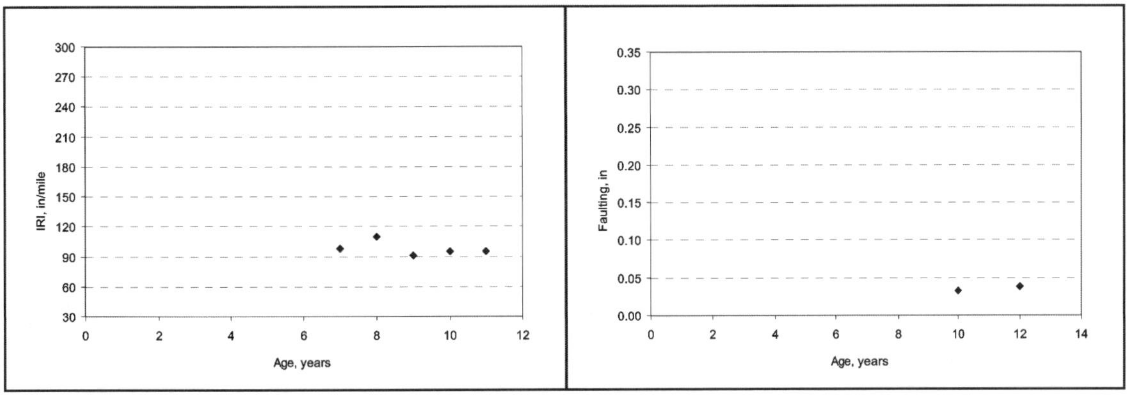

PMS site—E02-S-S2

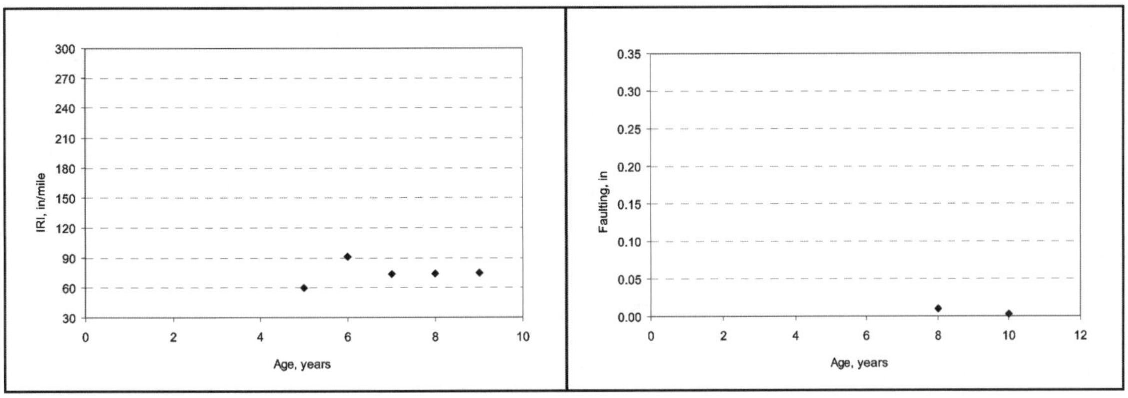

PMS site—E03-S-S1

Figure A3.2.B-1.—*Continued*

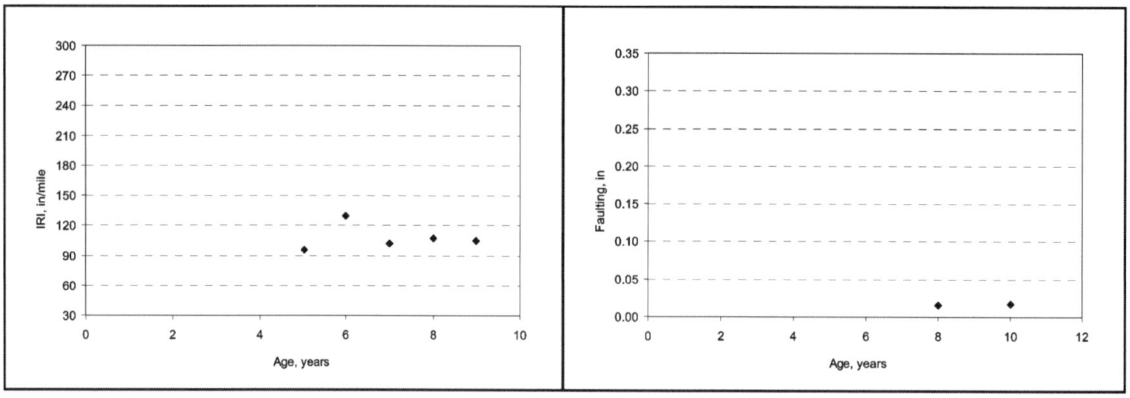

PMS site—E03-S-S2

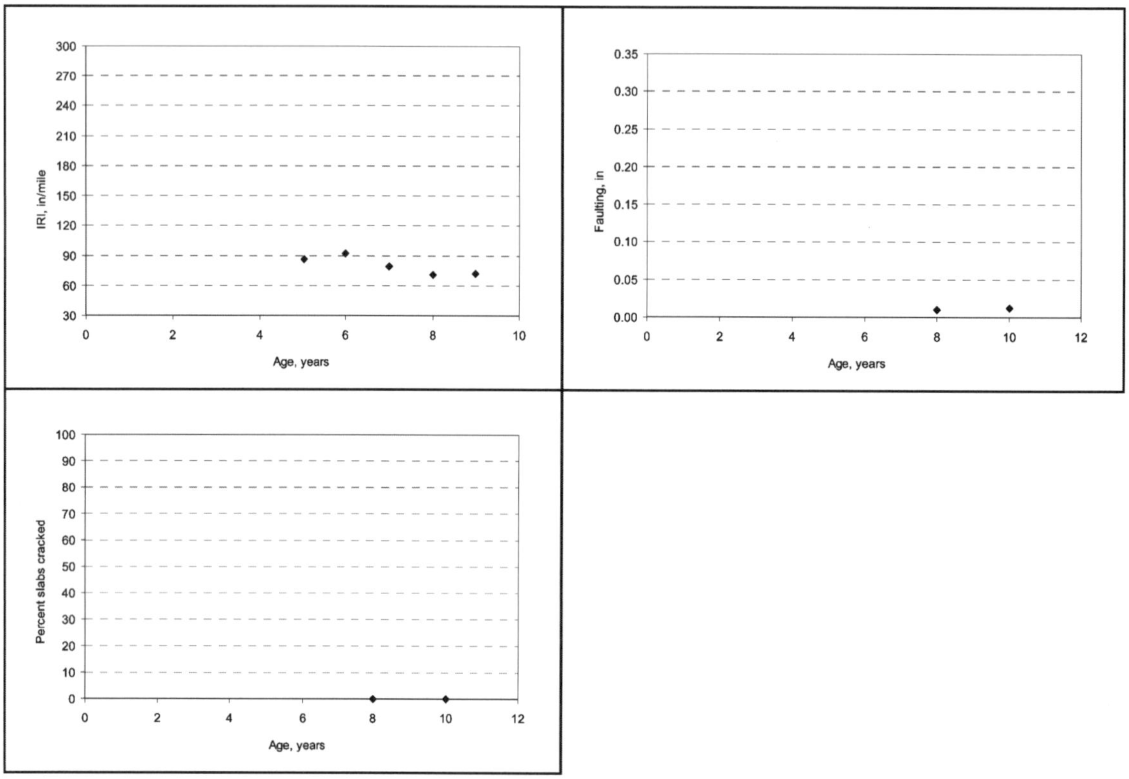

PMS site—E04-S-S1

Figure A3.2.B-1.—*Continued*

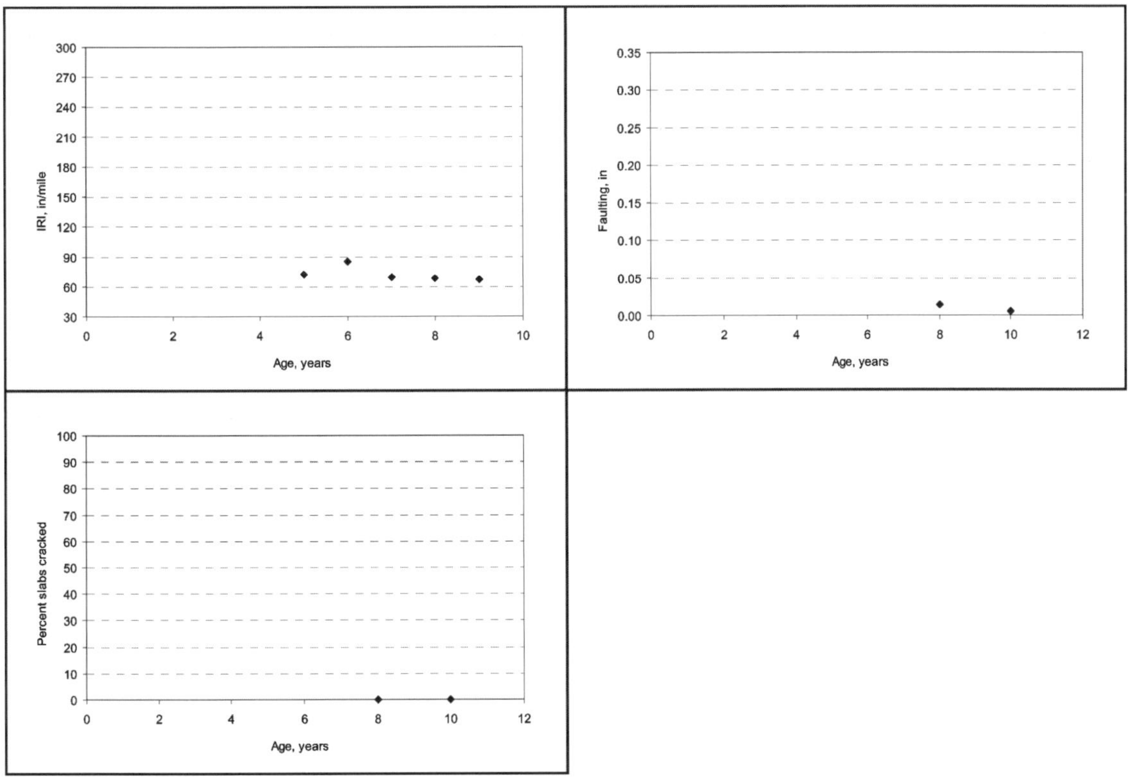

PMS site—E04-S-S2

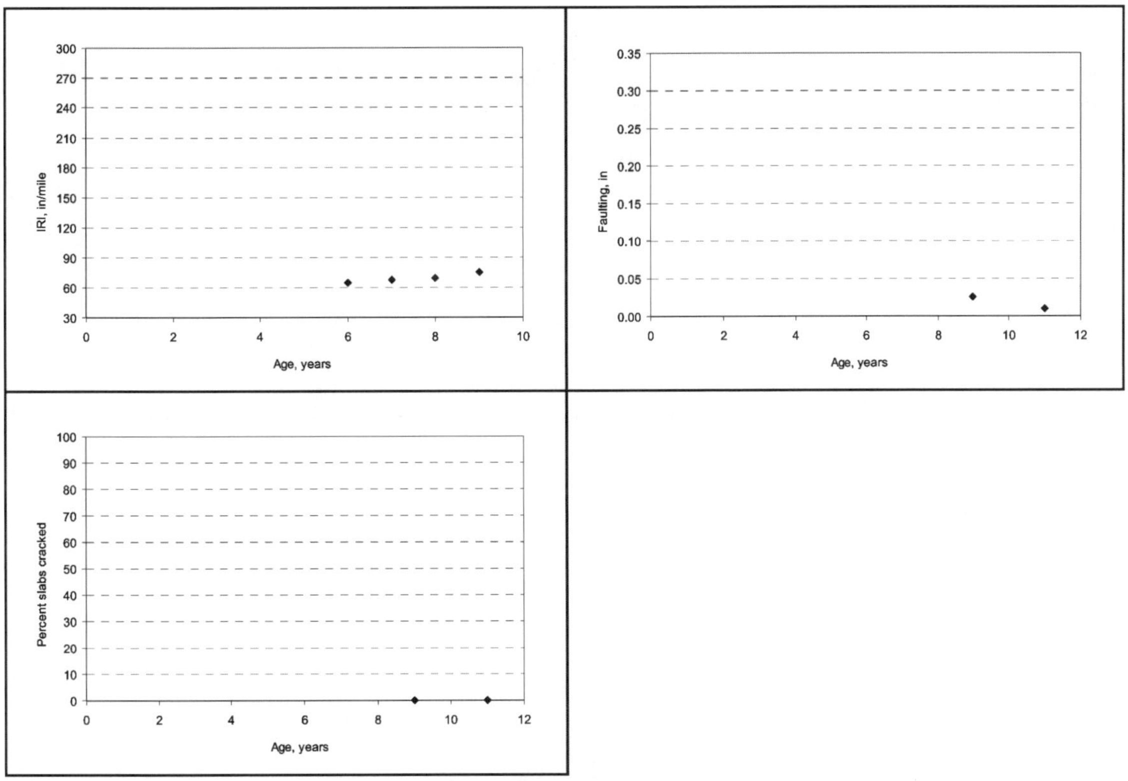

PMS site—F01-W-S1

Figure A3.2.B-1.—*Continued*

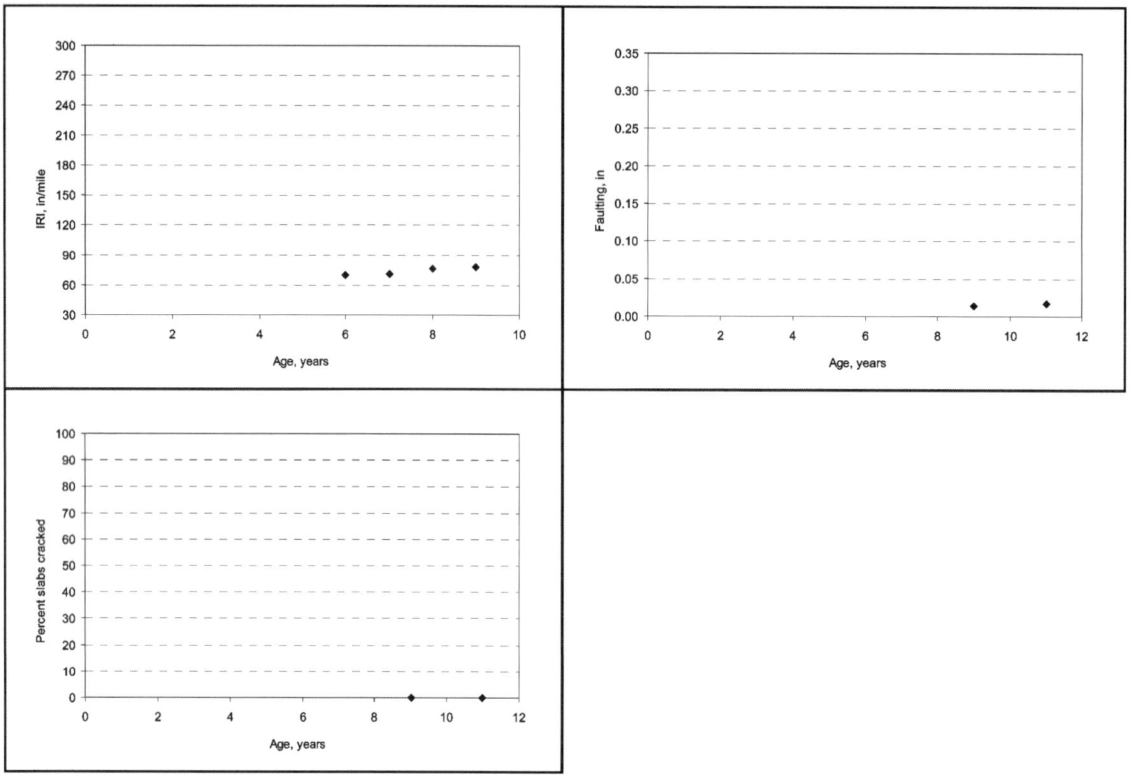

PMS site—F01-W-S4

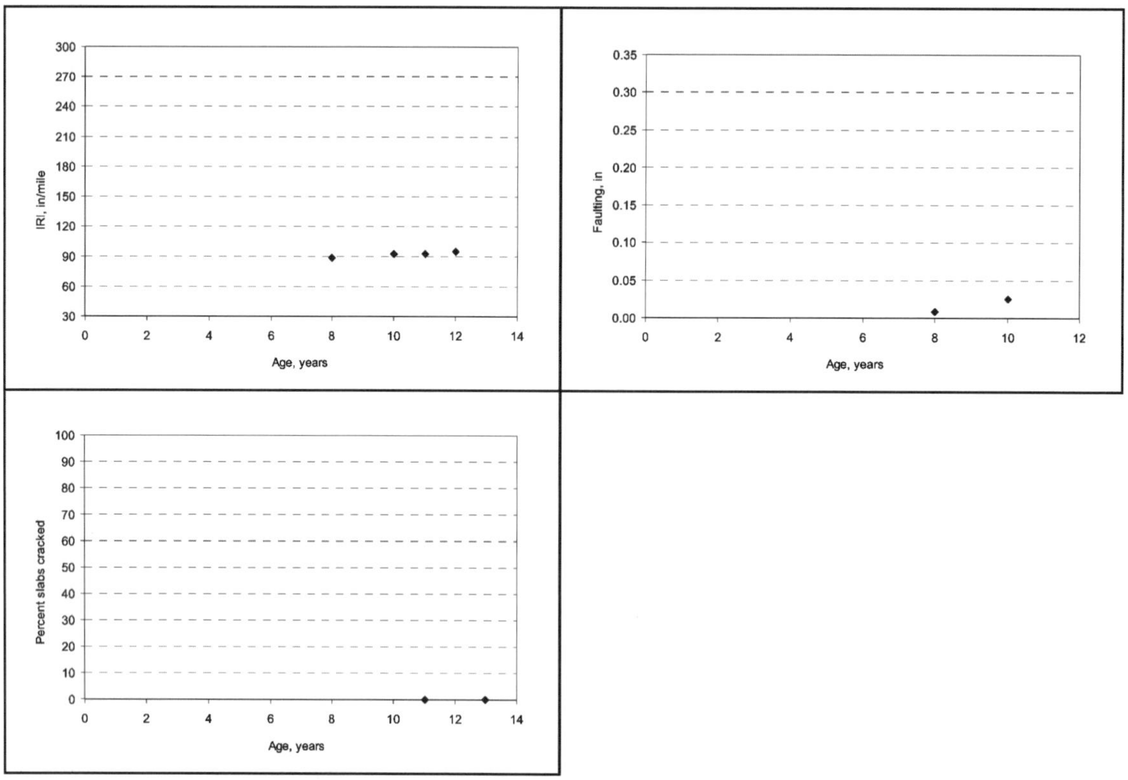

PMS site—F02-E-S1

Figure A3.2.B-1.—*Continued*

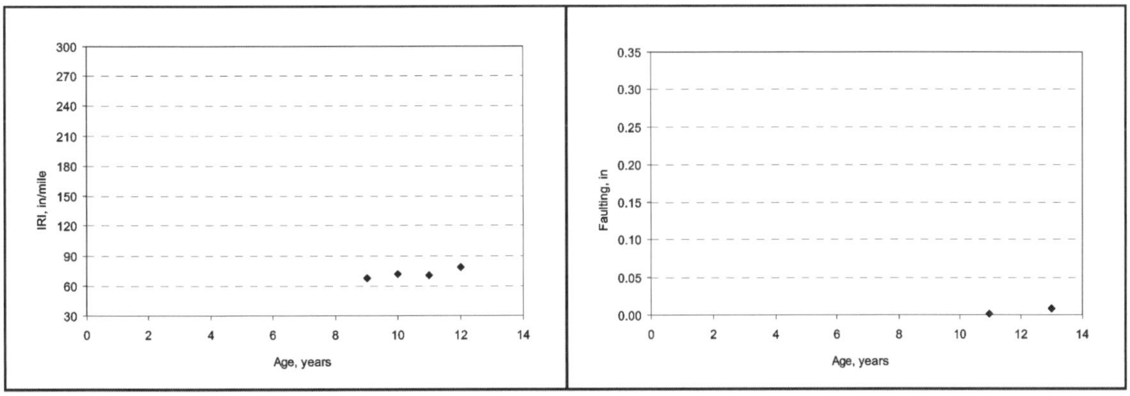

PMS site—F03-W-S1

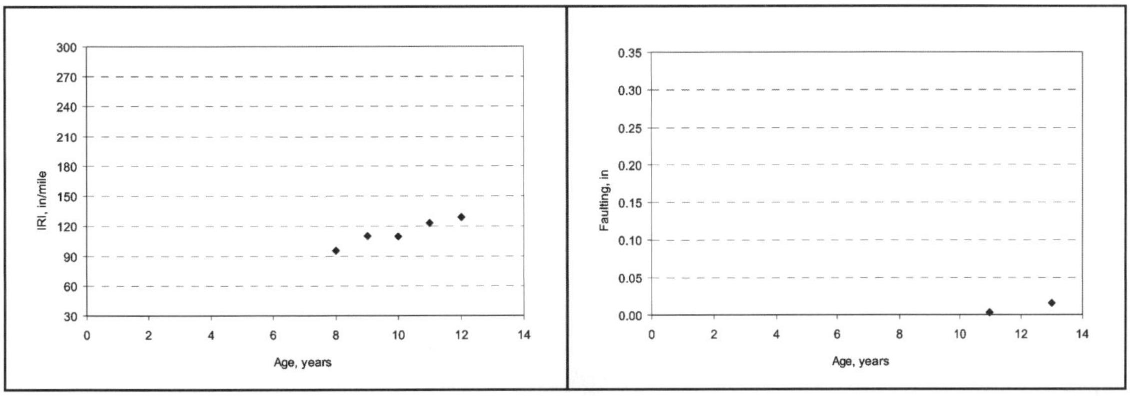

PMS site—F03-W-S2

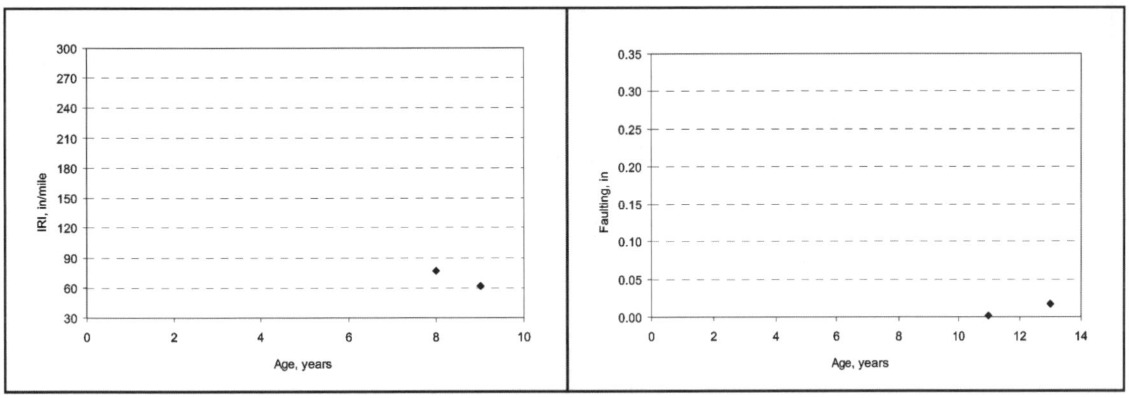

PMS site—F03-W-S3

Figure A3.2.B-1.—*Continued*

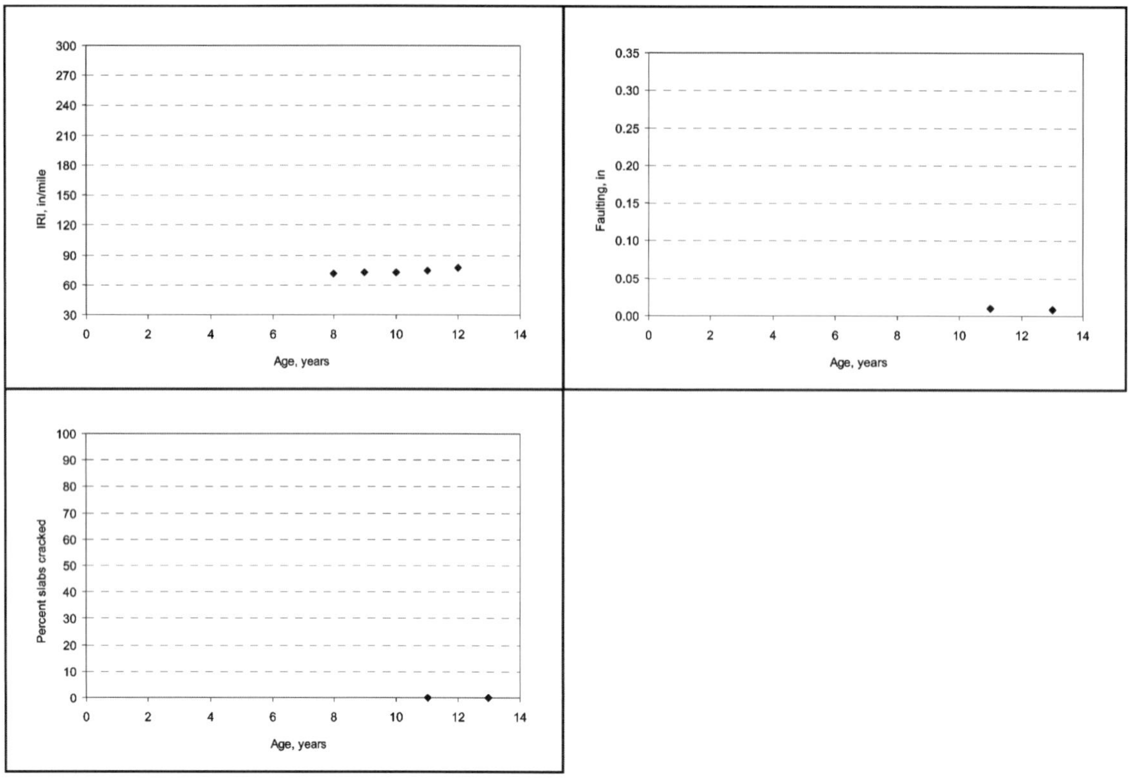

PMS site—F04-W-S1

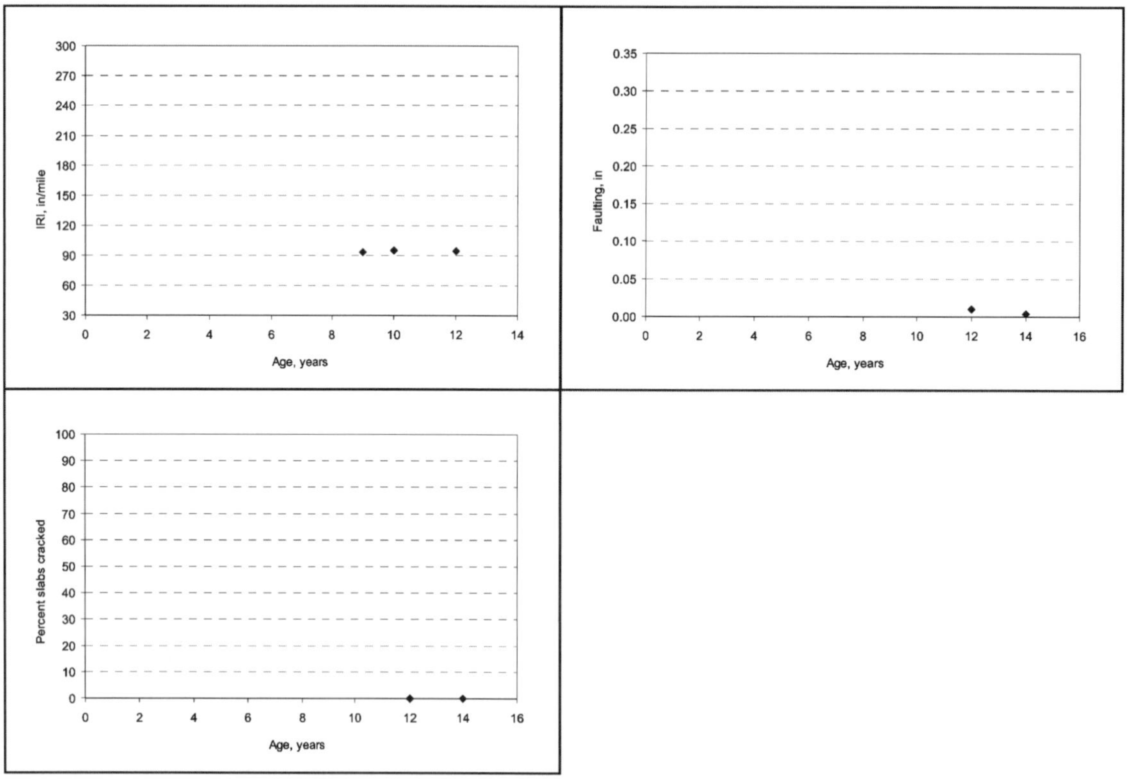

PMS site—F05-E-S1

Figure A3.2.B-1.—*Continued*

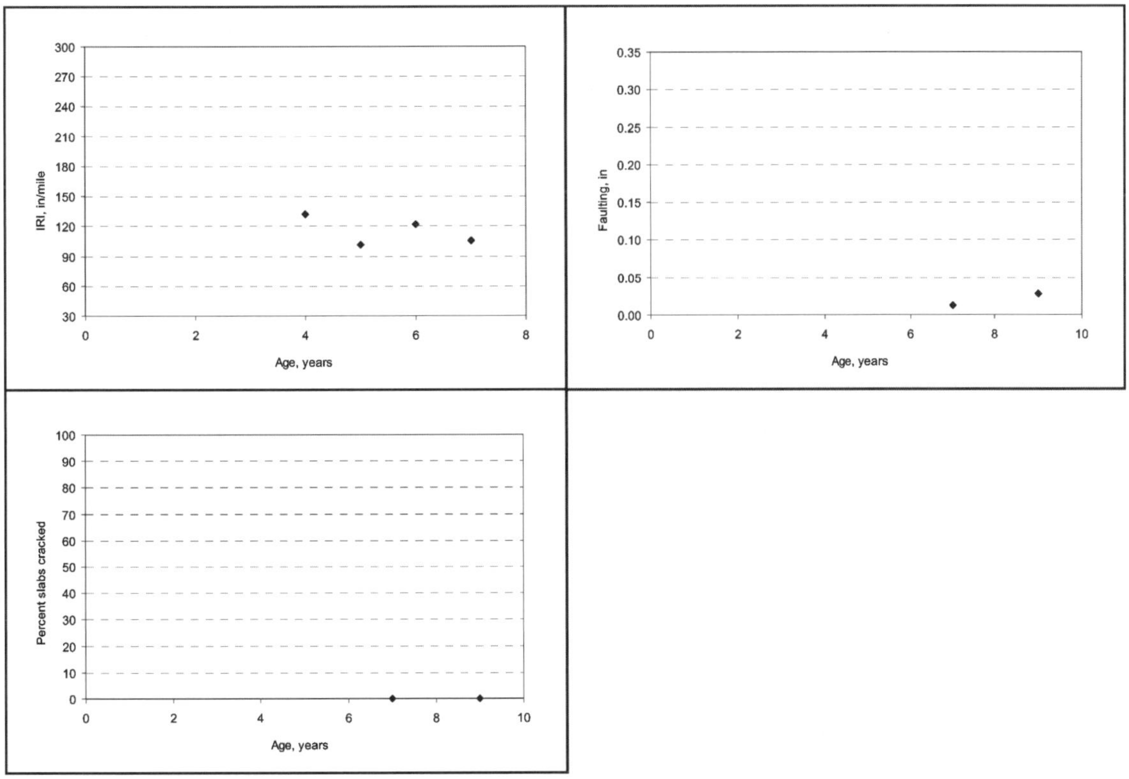

PMS site—G03-E-S1

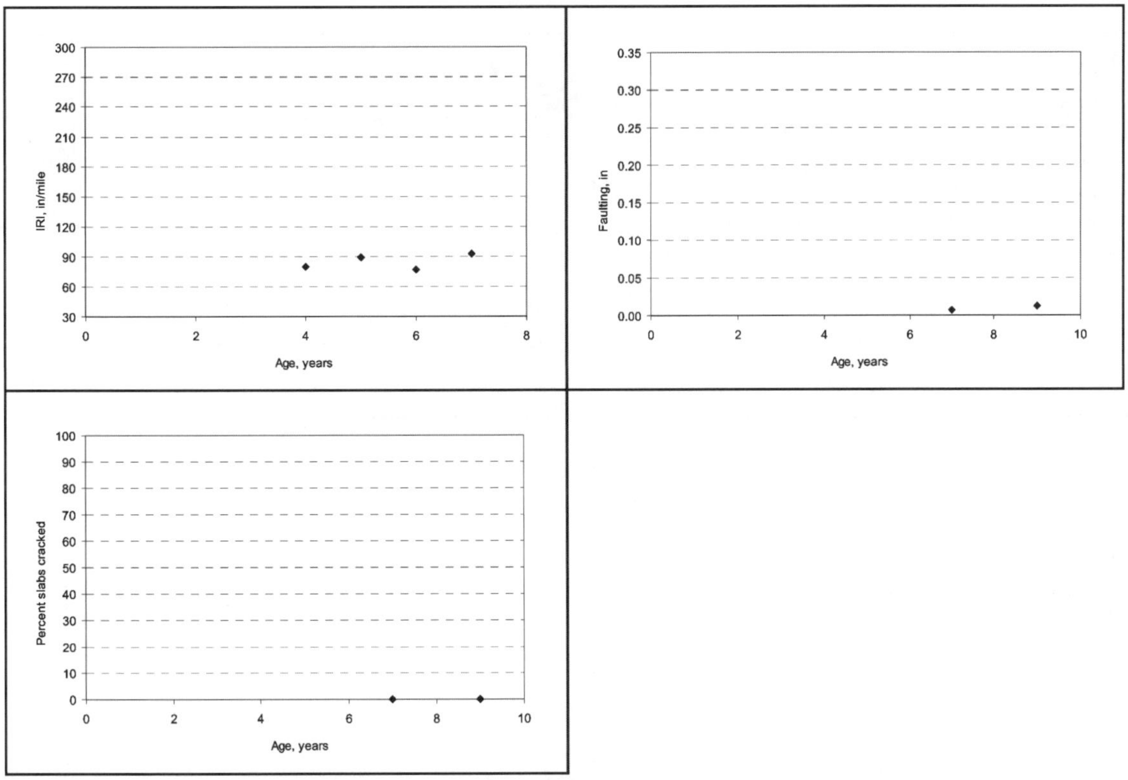

PMS site—G03-E-S2

Figure A3.2.B-1.—*Continued*

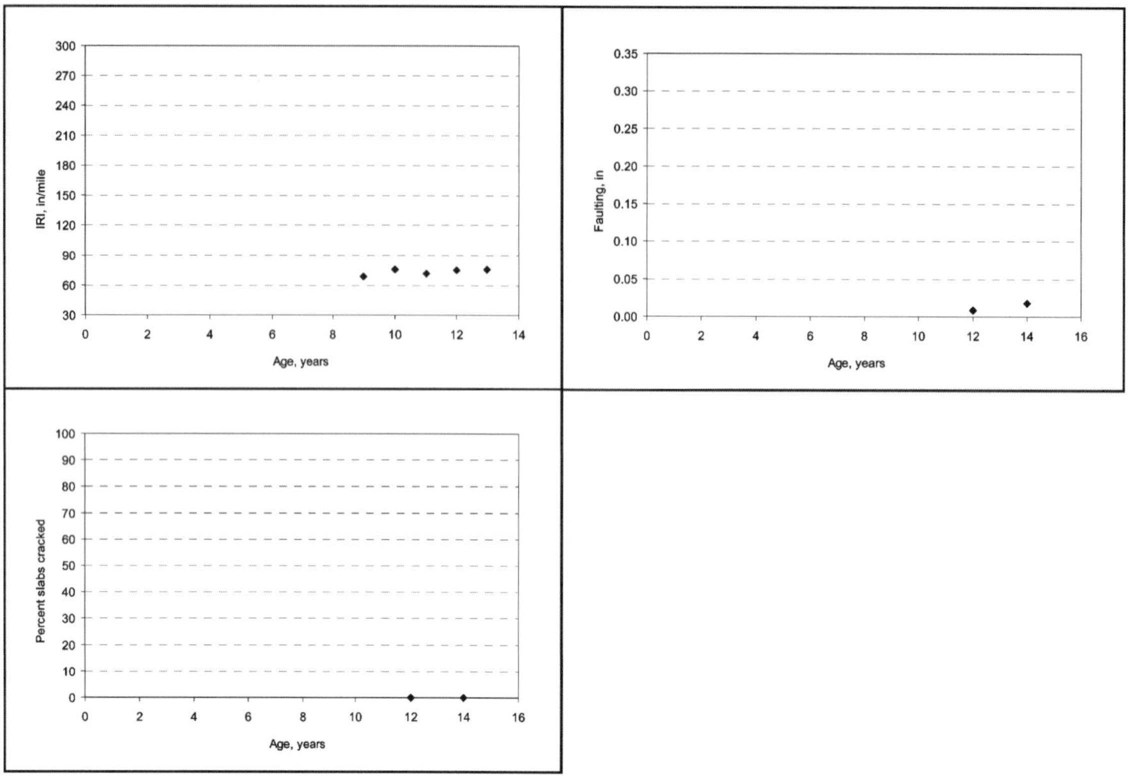

PMS site—F05-E-S2

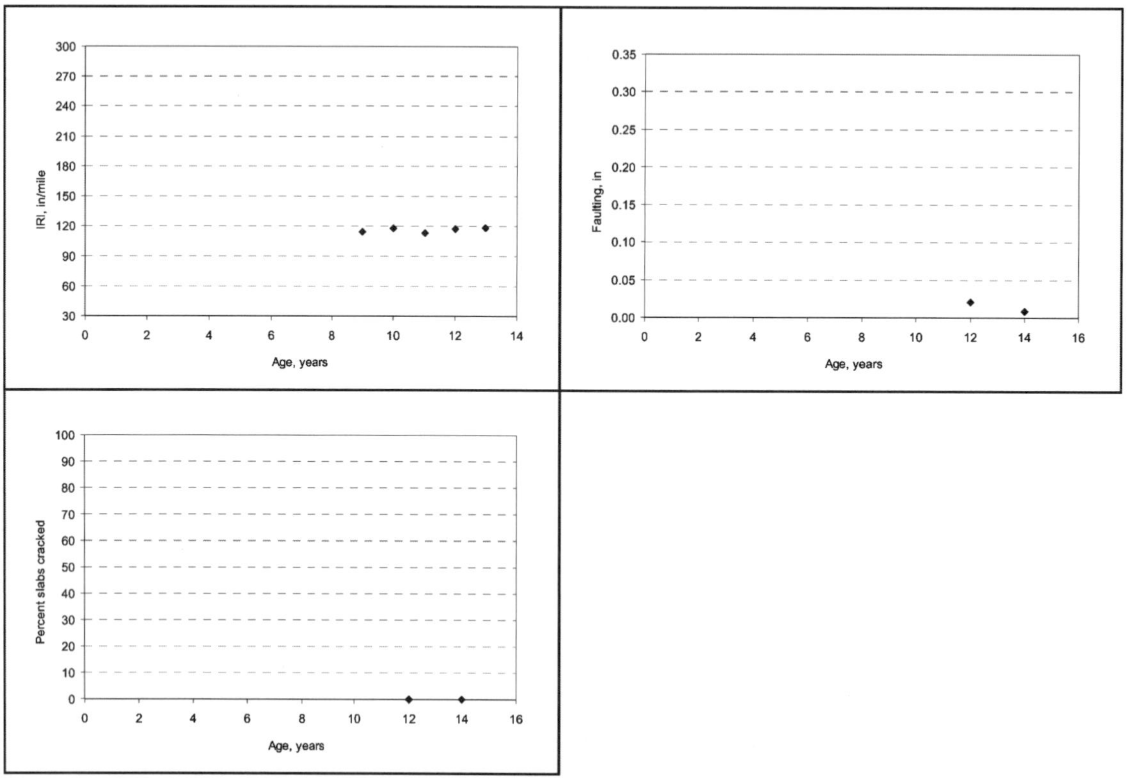

PMS site—F05-E-S3

Figure A3.2.B-1.—*Continued*

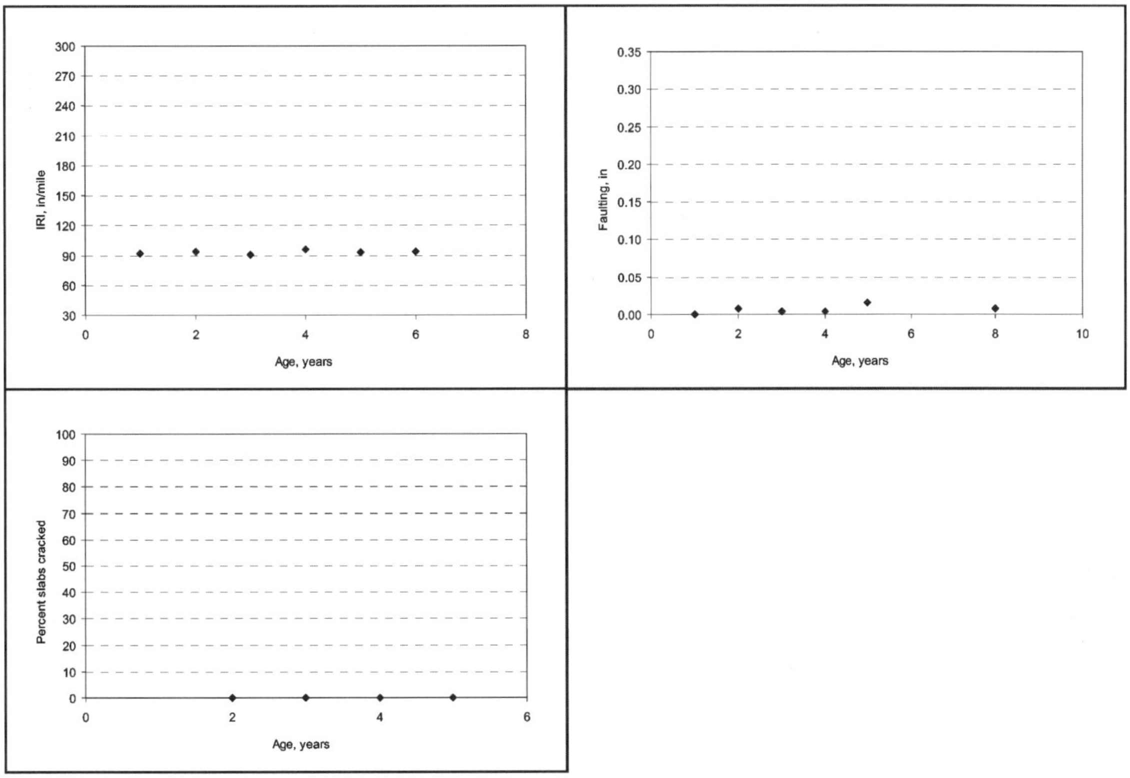

LTPP site—29_0807

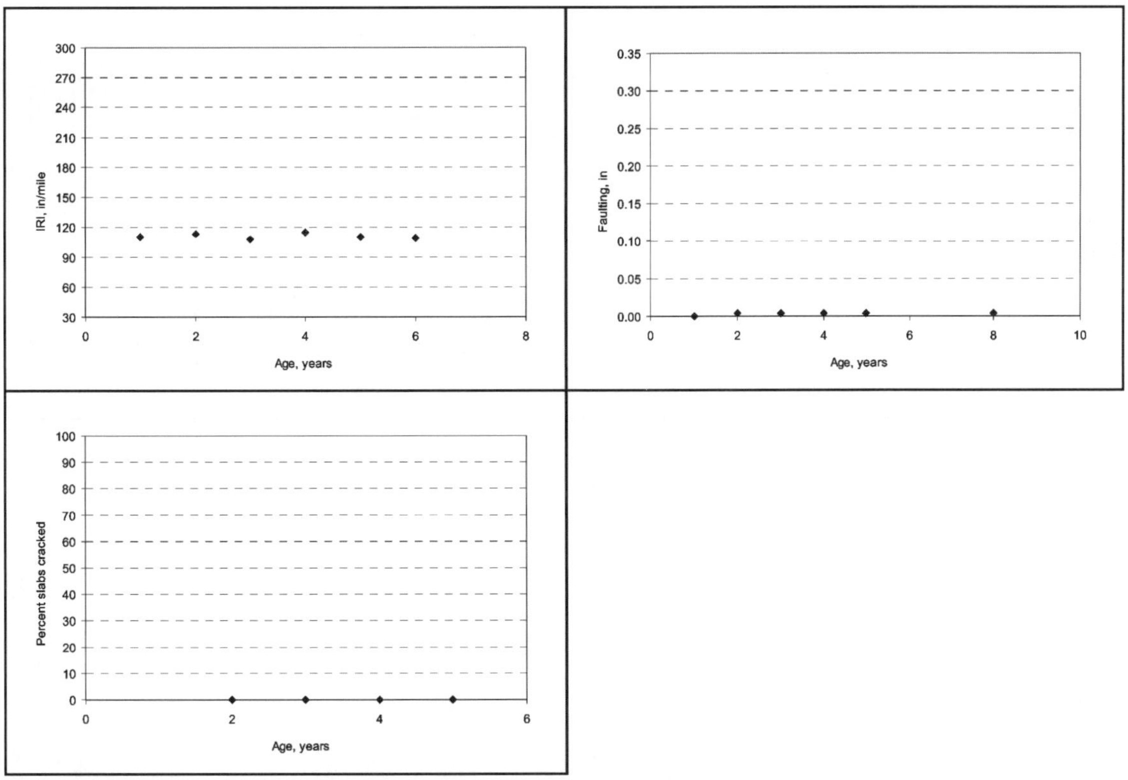

LTPP site—29_0808

Figure A3.2.B-1.—*Continued*

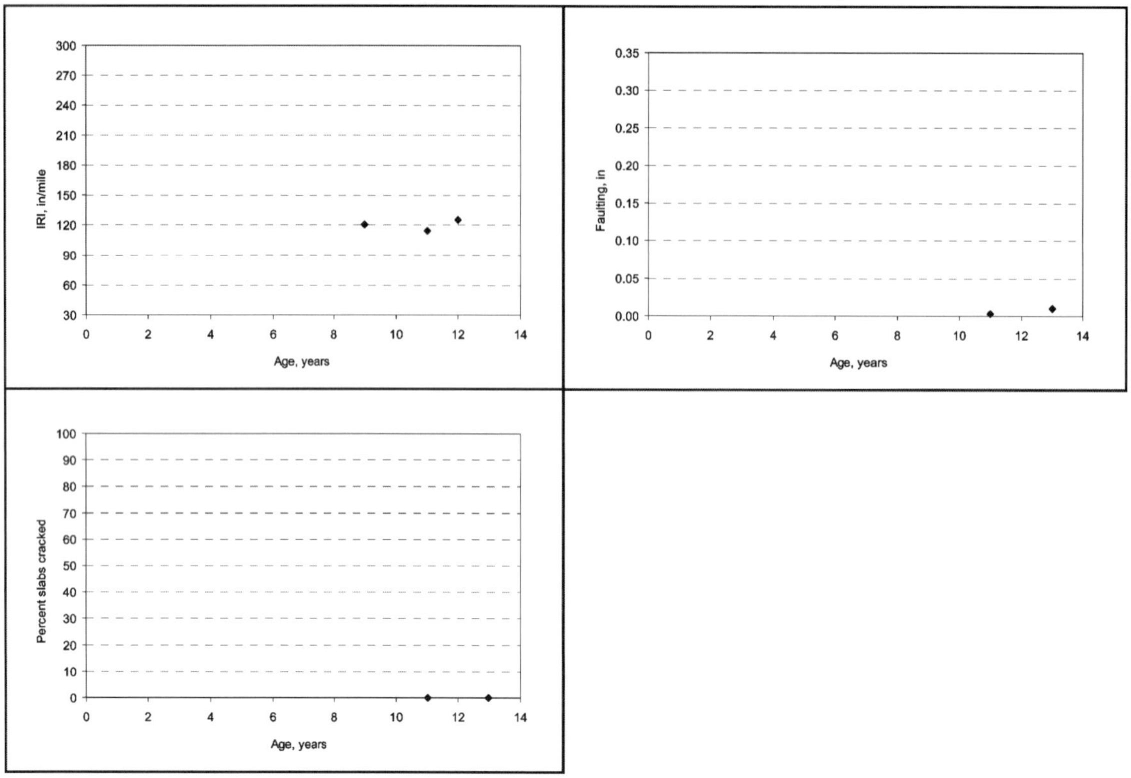

PMS site—F06-E-S1

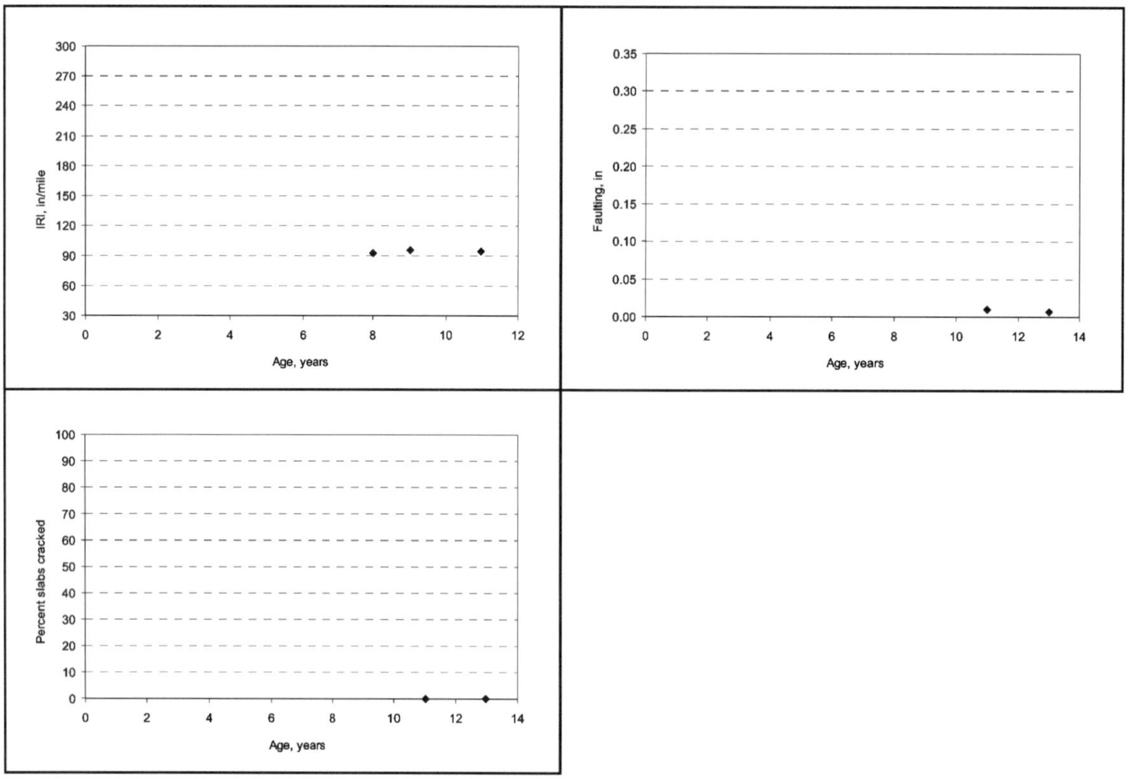

PMS site—F06-E-S2

Figure A3.2.B-1.—*Continued*

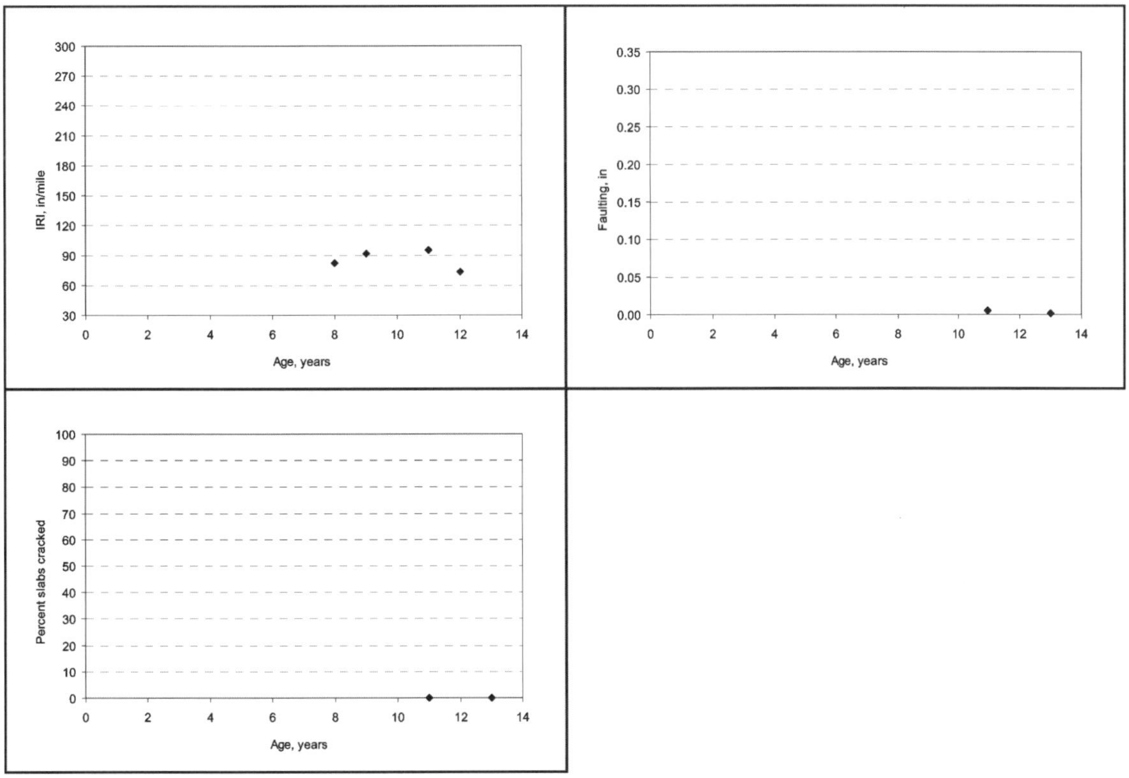

PMS site—F06-E-S3

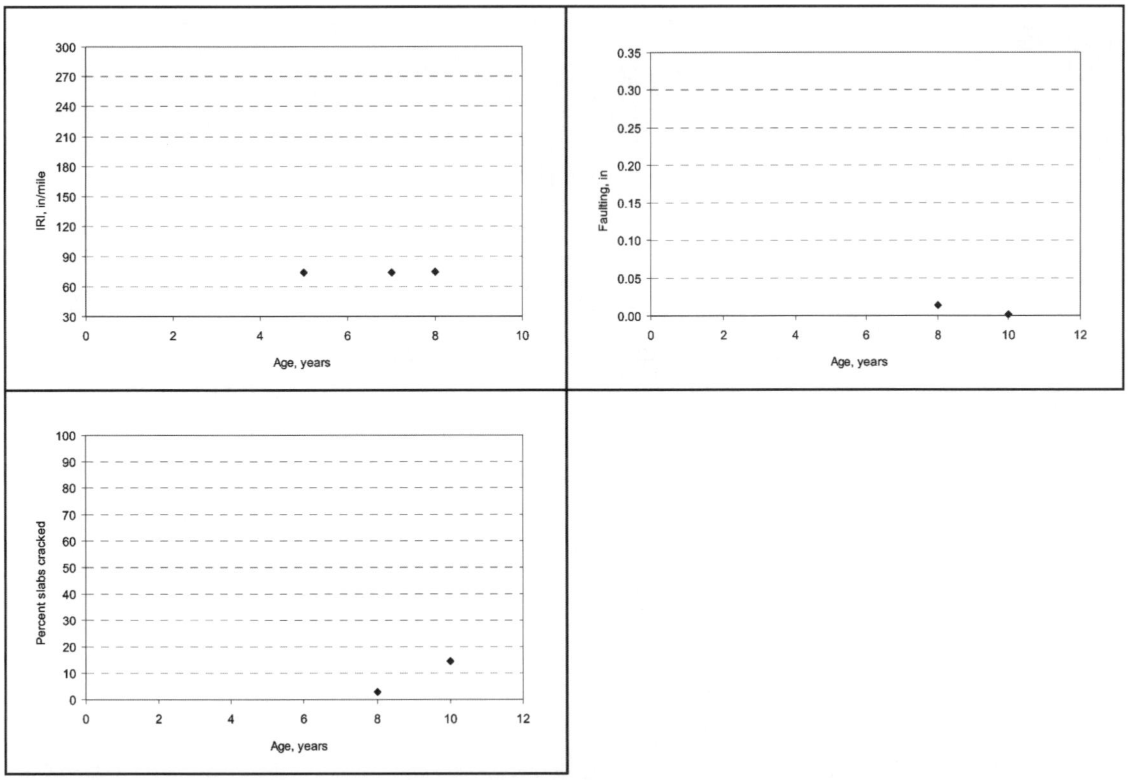

PMS site—F07-S-S1

Figure A3.2.B-1.—*Continued*

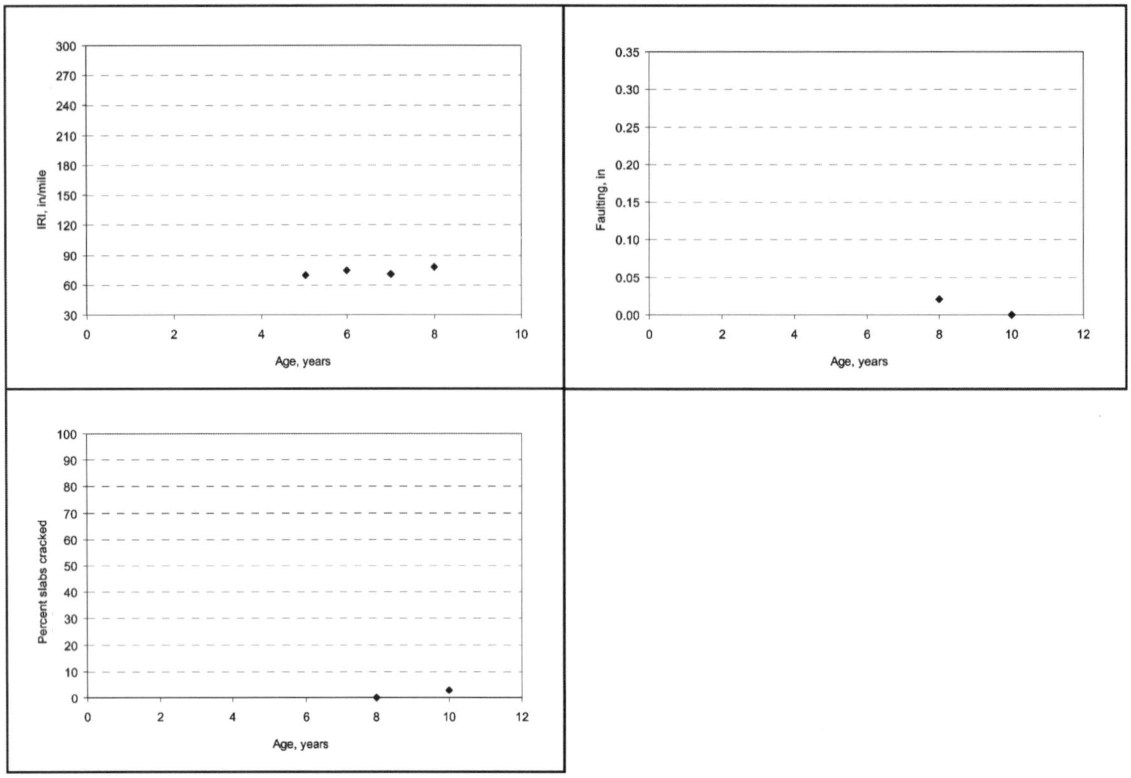

PMS site—F07-S-S2

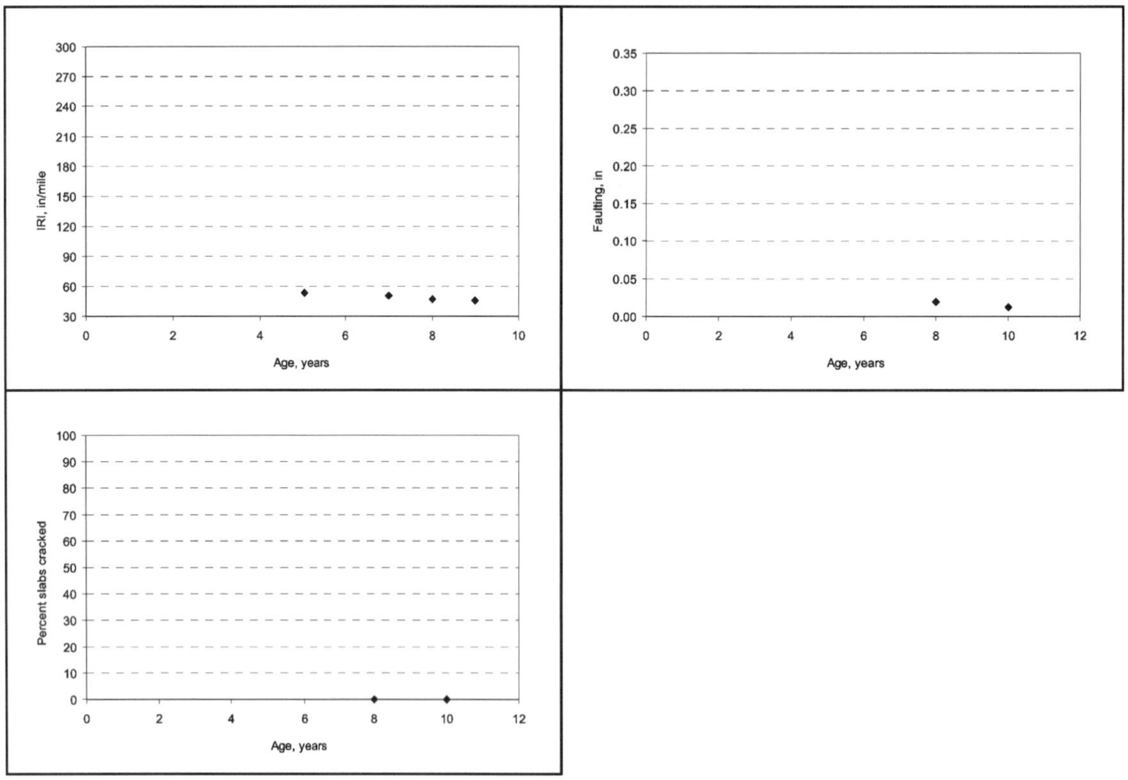

PMS site—F08-E-S1

Figure A3.2.B-1.—*Continued*

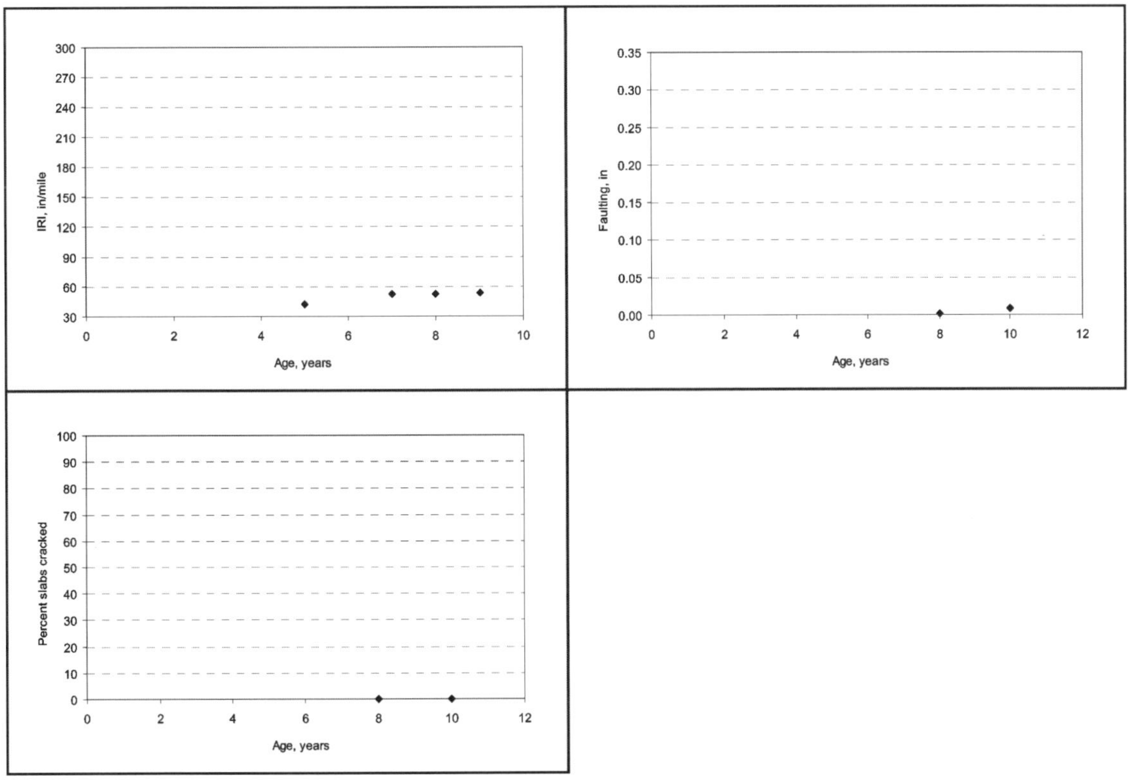

PMS site—F08-E-S2

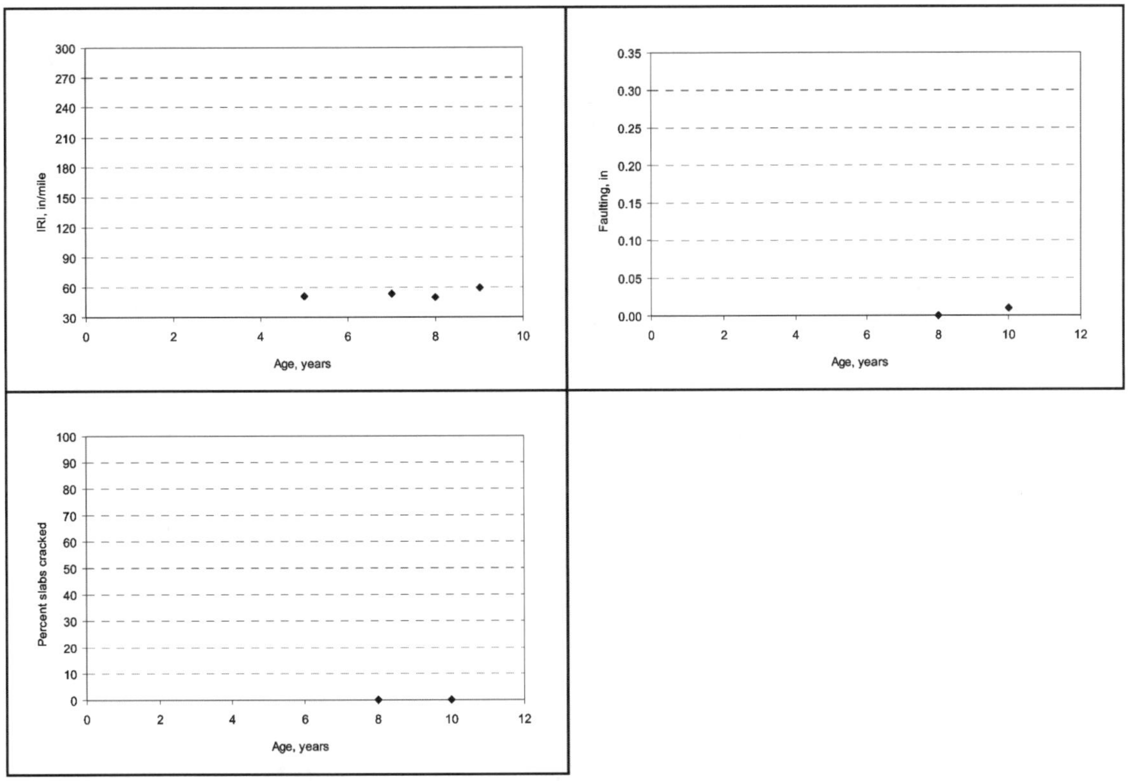

PMS site—F09-W-S1

Figure A3.2.B-1.—*Continued*

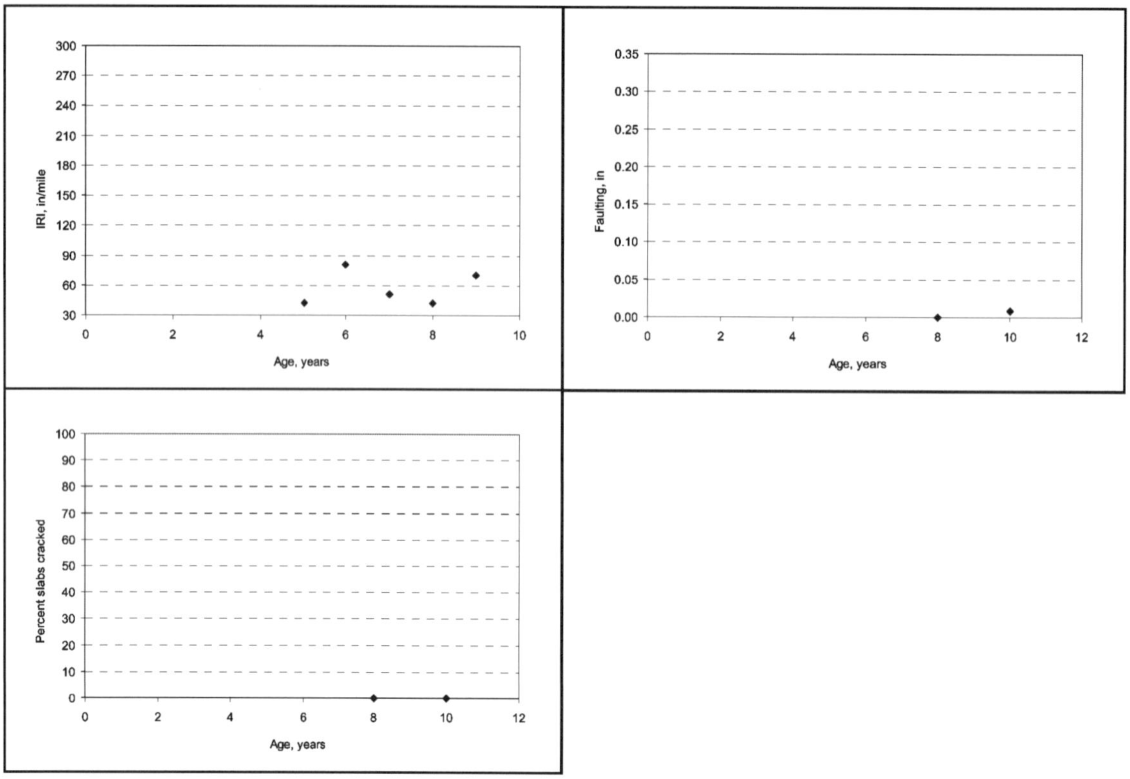

PMS site—F09-W-S2

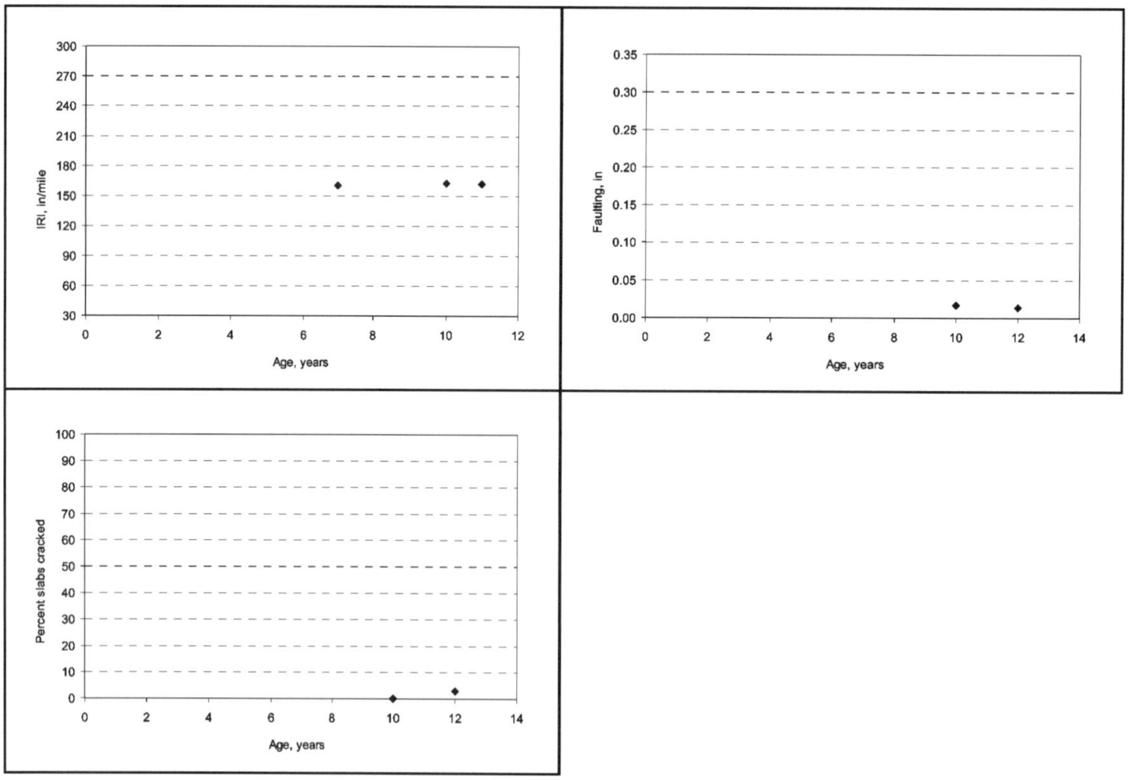

PMS site—F10-E-S1

Figure A3.2.B-1.—*Continued*

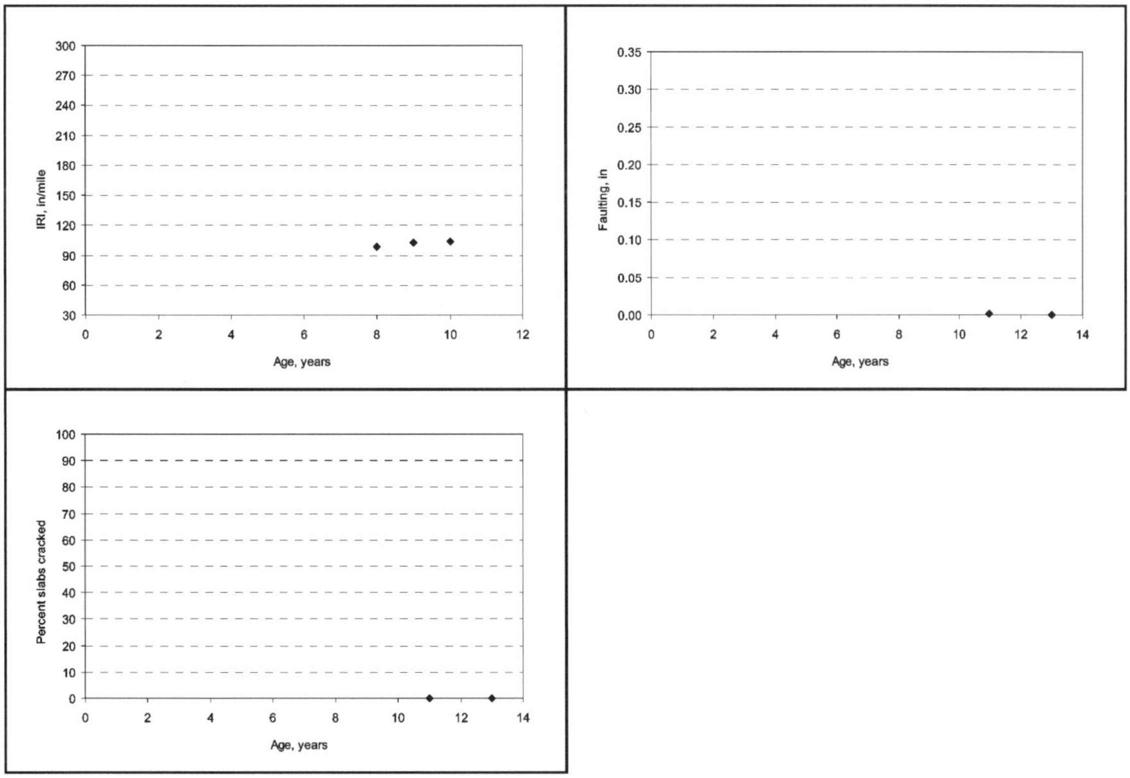

PMS site—G01-S-S1

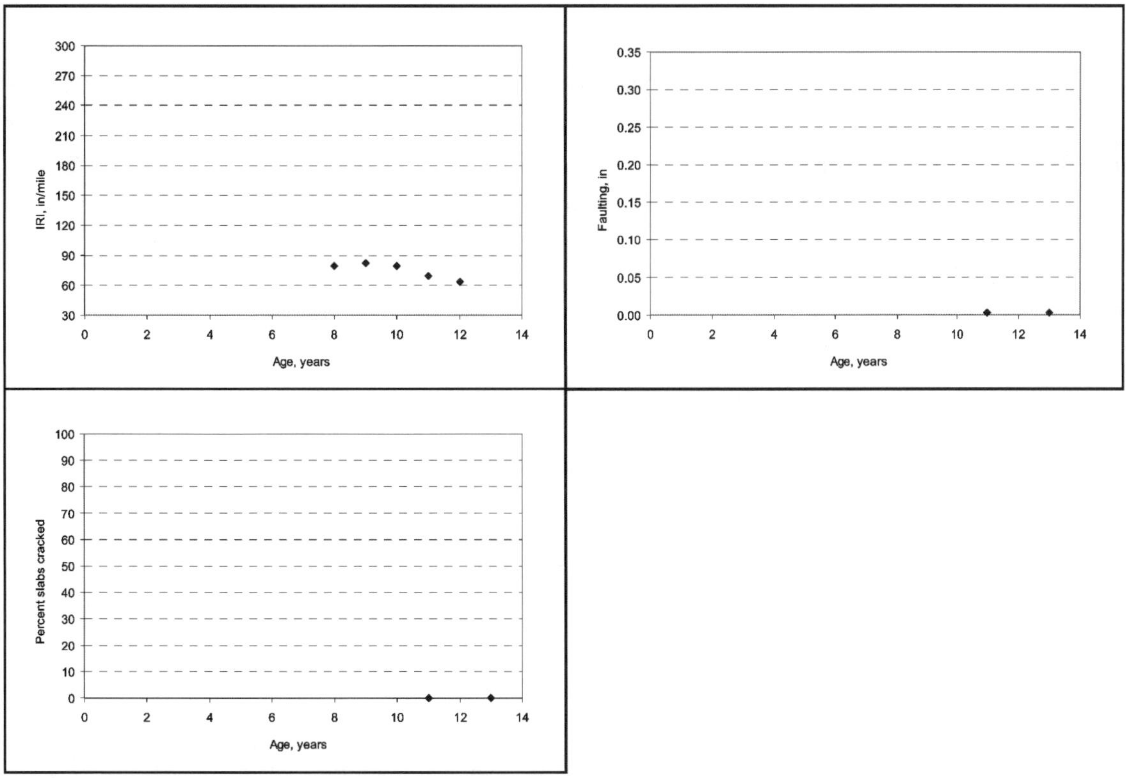

PMS site—G01-S-S2

Figure A3.2.B-1.—*Continued*

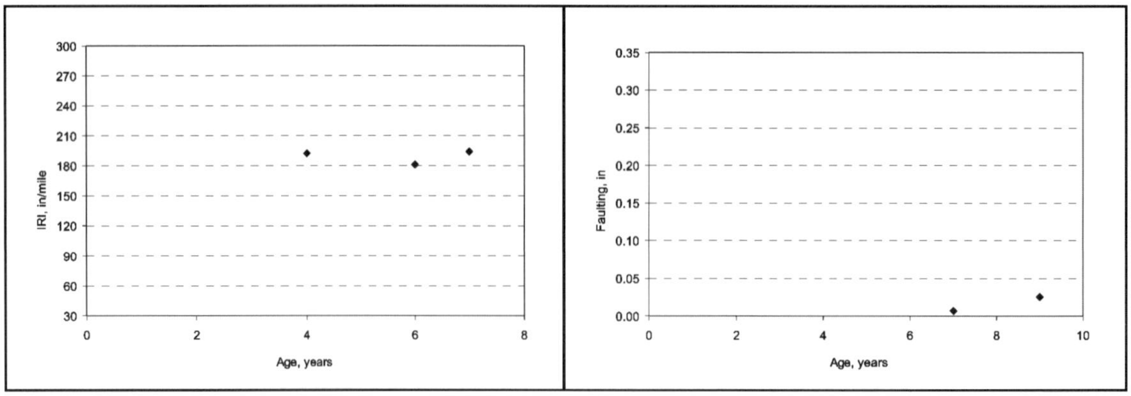

PMS site—G02-E-S1

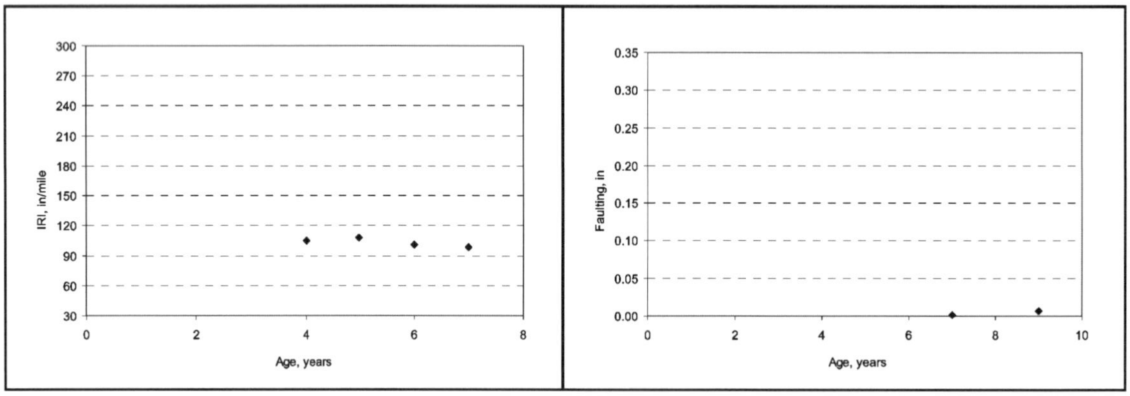

PMS site—G02-E-S2

Figure A3.2.B-1.—*Continued*

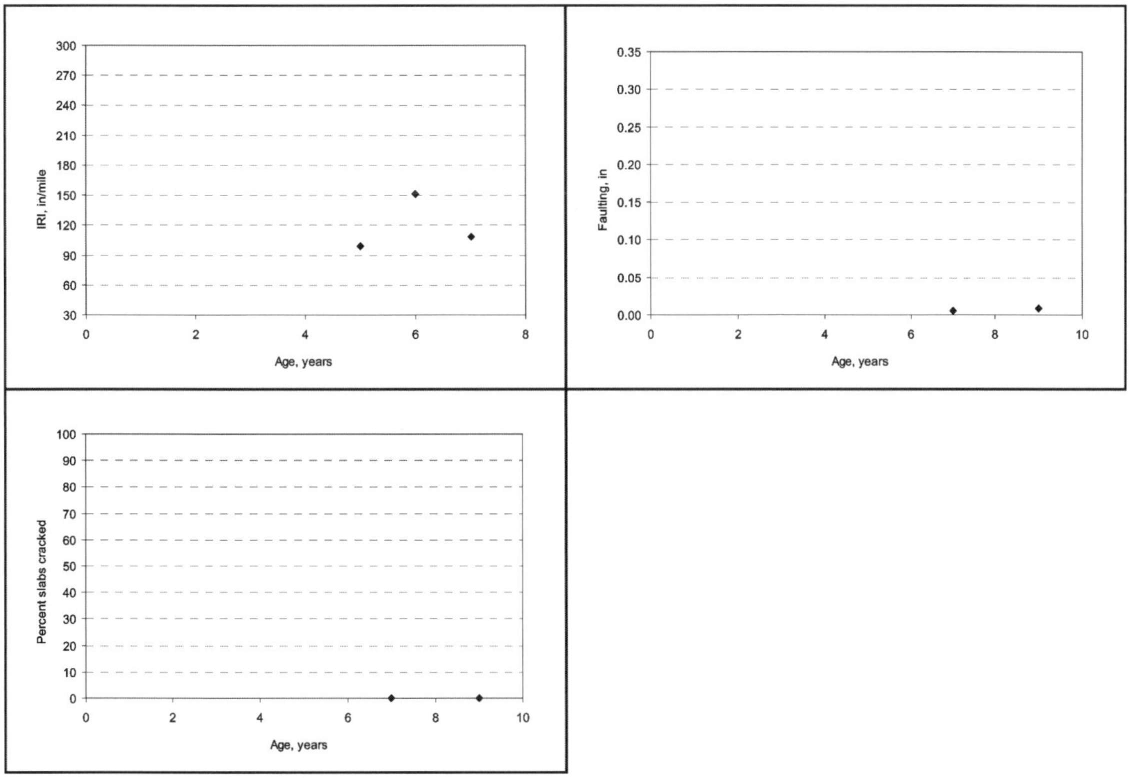

PMS site—G03-E-S3

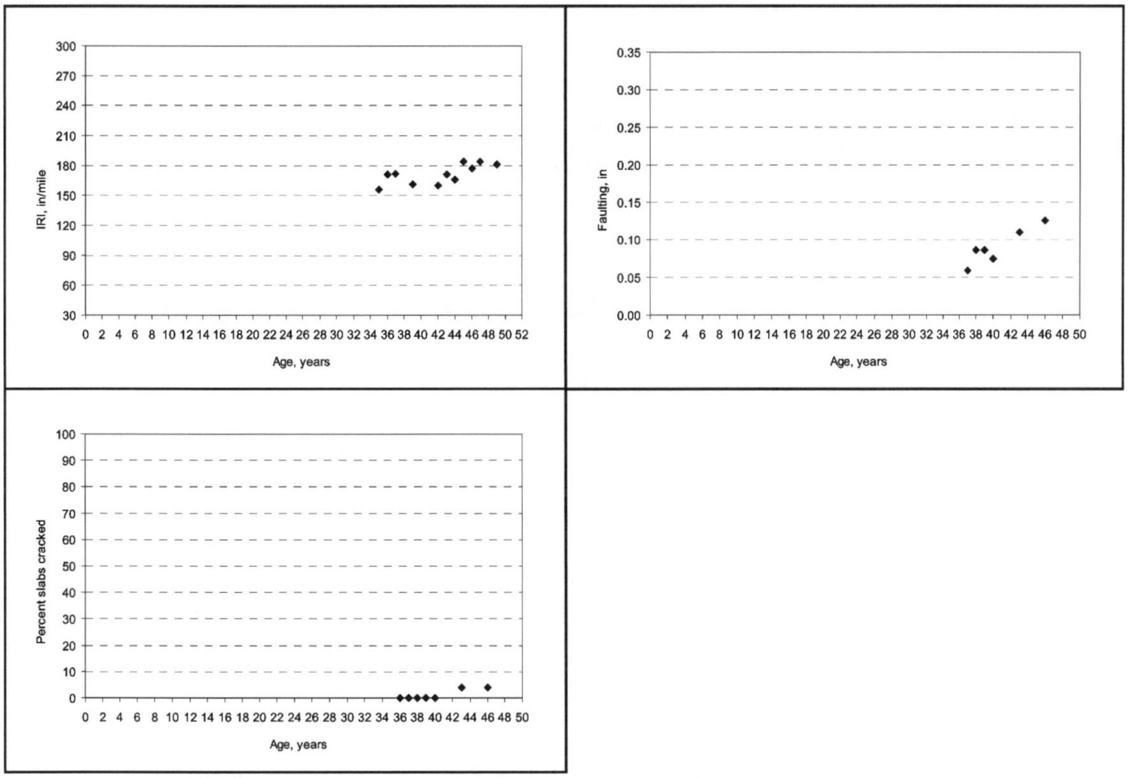

LTPP site—29_0701

Figure A3.2.B-1.—*Continued*

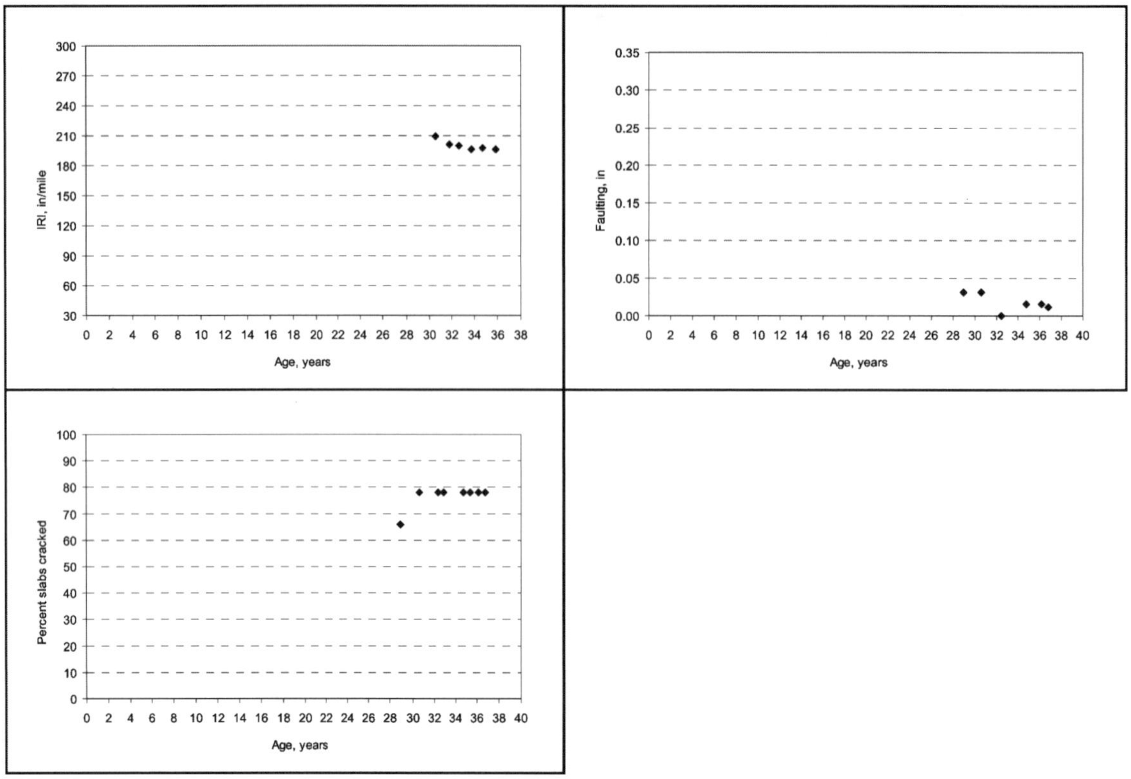

LTPP site—29_A601

Figure A3.2.B-1.—*Continued*

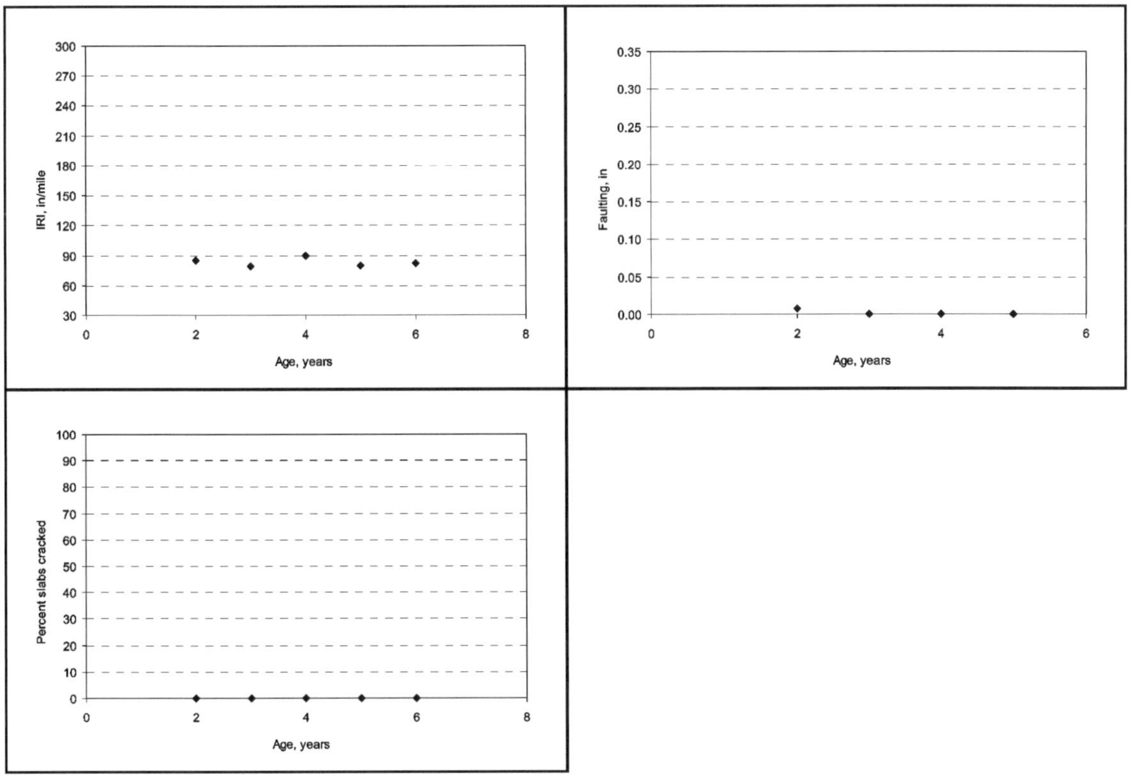

LTPP site—29_A807

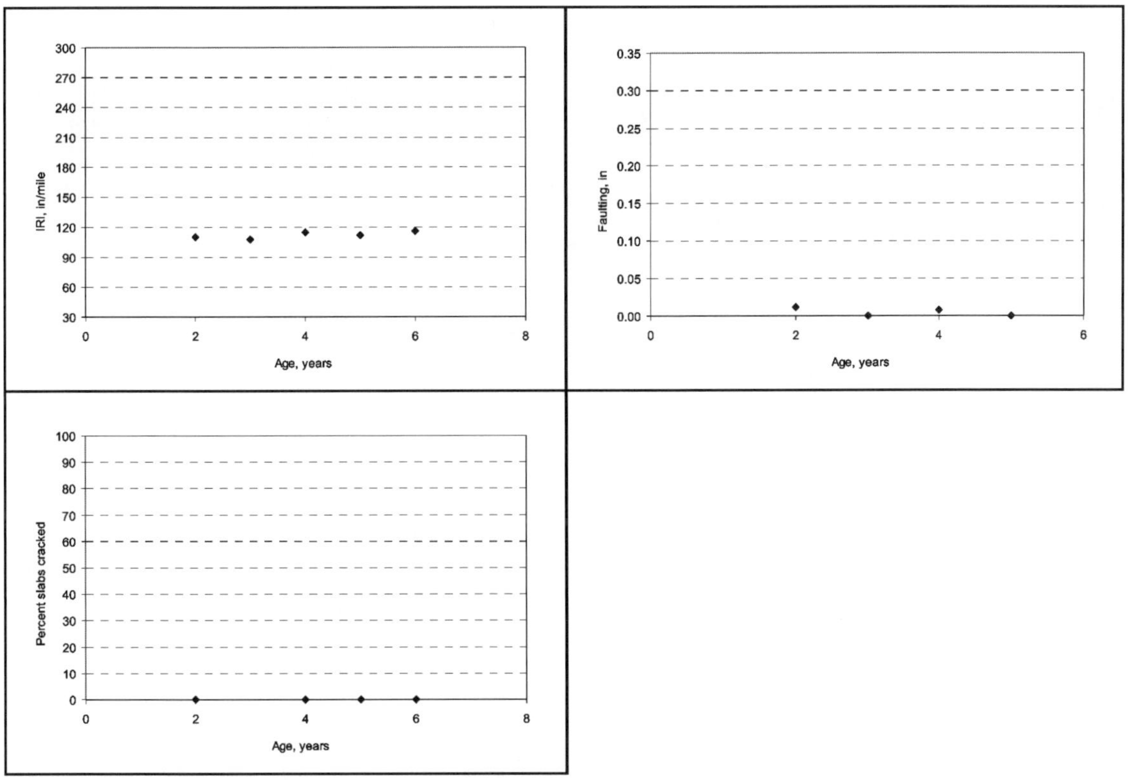

LTPP site—29_A808

Figure A3.2.B-1.—*Continued*